Unreal Engine 5 Game Development with C++ Scripting

Become a professional game developer
and create fully functional, high-quality games

Zhenyu George Li

BIRMINGHAM—MUMBAI

Unreal Engine 5 Game Development with C++ Scripting

Group Product Manager: Rohit Rajkumar

Publishing Product Manager: Vaideeshwari Muralikrishnan

Senior Editor: Hayden Edwards

Technical Editor: Simran Udasi

Copy Editor: Safis Editing

Project Coordinator: Aishwarya Mohan

Proofreader: Safis Editing

Indexer: Manju Arasan

Production Designer: Ponraj Dhandapani

Marketing Coordinators: Namita Velgekar & Nivedita Pandey

First published: August 2023

Production reference: 2150425

Published by Packt Publishing Ltd.

Grosvenor House

11 St Paul's Square

Birmingham

B3 1R

ISBN 978-1-80461-393-1

www.packtpub.com

Foreword

I have known and worked with George Li in various capacities for more than 20 years.

We met first as colleagues at a private college of interactive arts in downtown Vancouver, Canada. I was a full professor of linguistics at a university in the area, with extensive experience in developing a cognitive science program at the university. Establishing a curriculum for such a program involved familiarizing myself with areas well beyond the area of my specific academic position and collaborating closely with colleagues – particularly in computing science (cognitive), psychology, and philosophy (particularly concerning socio-epistemological issues such as AI). It was in this spirit that I was happy to associate myself with a private college, focusing on language study and teaching, together with preparing students to handle 3D computer applications (e.g., 3ds Max, 3D GameStudio, and Unreal) and showing them how to generate animation materials for film, TV, and games.

George Li was in charge of all the technical requirements of the college. I very quickly realized that he was not merely extremely competent and forthcoming (he had already occupied high-level computing-related positions, software engineering for instance, in China prior to emigrating to Canada) but also computationally competent and brilliantly innovative. He also had a particular interest in the development of game engines and was already collaborating with his colleague, Charles Yeh, on a practical reference book, *XNA PC & Xbox360 C# Game Programming*, with proprietary-produced text and games to his credit.

As colleagues in the college, George and I found common, mutually strengthening interests. Eventually, he and I found ourselves in charge of creating a two-year interactive-arts program curriculum for the college.

My whole career, at universities in the UK and Canada, and as a member of the editorial boards of a major academic journal and a very influential series, has closely involved the evaluation of the intellectual quality of people's capabilities and work, and in my opinion, George Li's innate talents shine out clearly throughout his work and will continue to do so in the future. His ability to express his knowledge of the subject at hand is outstandingly demonstrated in all his work, performance, expressions, and character.

This book, like the one co-authored with Yeh, is designed for independent developers and company training, plus for reference after post-secondary education. The knowledge presented herein is most intelligently, clearly, and effectively presented so as to be as efficiently applicable and pedagogically effective as possible on any device or platform, producing high-quality games, accessories, and edits.

This volume will certainly stand the test of time and fulfill the majority of the needs of those working in the field of gaming. However, I am certain that George will make further, very crucial contributions to this topic.

Until then, this volume will serve you excellently, helping you enjoy and prosper with your future activities and products.

Dr. E. Wyn Roberts (M.A. Ph.D (Cantab.))

Emeritus Professor of Linguistics, Simon Fraser University

Contributors

About the author

Zhenyu George Li is a passionate video game developer with over 20 years of experience in the field. As a seasoned software engineer, George has contributed significantly to the development of numerous games throughout his career and currently serves as a senior development consultant at Unity.

George's fascination with video games was sparked during his college studies, igniting a passion that would shape his professional journey. During the early stages of his game development endeavors, George immersed himself in technologies such as Visual Basic, C/C++, DirectX, OpenGL, and Windows GUI. These foundational experiences laid the groundwork for his subsequent success in the industry.

Throughout his career, George has made substantial contributions to various commercial games. Notable titles in his portfolio include Halo Infinite, Magic: The Gathering Arena, Stela, Dead Rising 2, The Bigs 2, and so on. His involvement in these projects has allowed him to gain extensive knowledge and practical experience in a wide range of domains, including programming, game engines, gameplay and AI, graphics, animation, multiplayer games, multiplayform games, and game physics. In practical applications, George has used both the Unreal and Unity engines in the development of real game projects.

In addition to his achievements as a game developer, George has also honed his teaching abilities during his eight years of college-level instruction. He has shared his knowledge and expertise with aspiring developers, serving as a lecturer at the Vancouver Film School (VFS), the College of Interactive Arts, and Hefei Union University. During his teaching at VFS, George instructed students in the intricacies of Unreal Engine.

I express my gratitude to my wife, Alison Guo, for her support in handling family responsibilities and for enabling me to dedicate time to completing this book. I also extend my thanks to Sarah Beck and Willy Campos for their encouragement and support throughout the writing process.

About the reviewers

Aditya Dutta holds a game programming advanced diploma from Humber College and is a highly accomplished senior software engineer at Archiact Interactive, bringing expertise in Unreal Engine and C++ system design and implementation.

With a strong collaborative spirit, Aditya leads feature development and takes ownership of tools while actively improving team processes. These leadership skills were evident during his tenure as lead programmer at Humber College, where he successfully guided and mentored a team of programmers, overseeing the technical aspects of significant projects in the virtual production and architecture industries.

His contributions at UP360 Inc. as a programmer included shipping numerous training VR simulations, developing iterative tools, and implementing gameplay mechanics using Unreal Engine.

Michael Oakes is a senior software consultant for Unity and has over 27 years of experience in the IT industry. He has worked with real-time 3D and games for over eight years, specializing in mixed reality design and development, shader programming, and AI and multiplayer systems.

He has worked as a technical consultant on other titles, including Packt Publishing's *Learn ML-Agents – Fundamentals of Unity Machine Learning*, written by Micheal Lanham.

Table of Contents

Part 1 – Getting Started with Unreal C++ Scripting

1

2

3

Learning C++ and Object-Oriented Programming   43

4

Investigating the Shooter Game's Generated Project and C++ Code   89

Part 2 – C++ Scripting for Unreal Engine

5

Learning How to Use UE Gameplay Framework Base Classes   117

6

Creating Game Actors   149

7

Controlling Characters   179

8

Handling Collisions   205

9

Part 3 – Making a Complete Multiplayer Game

10

11

Controlling the Game Flow   295

12

Polishing and Packaging the Game   327

Preface

Welcome, and thank you for choosing to pick up the *Unreal Engine 5 Game Development with C++ Scripting* book! This comprehensive book is designed to assist game developers and students in advancing their professional skills in C++ programming for Unreal Engine game development.

Unreal Engine is a powerful and versatile game engine widely used in both the gaming and movie-making industries. Possessing advanced and professional Unreal Engine development skills enables individuals to adapt more effectively to the demands of a career in game development, opening a multitude of opportunities for them.

When developing with Unreal Engine, you have the option to use either one or both of two available programming tools:

- **Blueprint** provides a user-friendly interface suitable for non-programmer developers
- **C++** is predominantly employed by software engineers, providing a more robust and flexible approach to game development

As an Unreal Engine developer, you may have a genuine interest in understanding C++ and how it integrates with the engine, even if you don't identify as a software engineer or aspire to become one.

This book is designed to assist you in expanding your knowledge and skills by guiding you through the necessary steps to create a fully fledged game, covering essential aspects of game development. It aims to smoothen the learning curve, allowing for a more seamless and efficient grasp of the concepts presented. The carefully organized topics eliminate the need for random searching and prevent wasted time on unrelated readings, enabling you to focus on the relevant information. Moreover, this book serves as a valuable reference manual, offering a comprehensive resource that can be revisited and utilized for further study.

Who this book is for

It is important to note that this book does not serve as a beginner's guide to using Unreal Engine. Prior to exploring its contents, you should already possess a basic understanding of, and practical experience with, Unreal Engine and Blueprint. This prerequisite ensures that you have a solid foundation of knowledge to fully leverage the material covered in this book, maximizing your learning experience.

This book caters to a diverse range of readers:

- Non-engineer game developers, such as game designers and artists who aspire to learn and comprehend C++ in the context of Unreal Engine development

- Software engineers who may lack prior experience in Unreal Engine C++ programming but wish to quickly acquire the necessary skills for their next project or job

- Students who are interested in learning and digging into Unreal C++ programming for their study or personal projects

- Individuals with a keen interest in game development using Unreal Engine will benefit from the comprehensive knowledge presented within these pages

What this book covers

Chapter 1, Creating Your First Unreal C++ Game, guides you quickly through creating a new C++ game project based on the Shooter template in Unreal. This chapter also introduces how to convert an existing Blueprint game project into a C++ game project.

Chapter 2, Editing C++ Code in Visual Studio, provides basic information on how to use the powerful integrated development environment Microsoft Visual Studio to edit C++ code. This chapter not only presents the editing skills needed but also demonstrates how to create a calculator application in C++.

Chapter 3, Learning C++ and Object-Oriented Programming, goes deeper into C++ programming based on the previous chapter's calculator project. This chapter covers the fundamental C++ syntax, data types, flow control, and so on. C++ object-oriented programming is also introduced in this chapter.

Chapter 4, Investigating the Shooter Game's Generated Project and C++ Code, explores the details of the generated shooter game project, including the project files' structure and the source files. In this chapter, the C++ code lines are briefly explained, so that you gain an overall understanding of how C++ code works.

Chapter 5, Learning How to Use the UE Gameplay Framework Base Classes, instructs you on how to create our new top-down game project, *Pangaea*. You will be guided on how to create the game actors, `DefenseTower`, for instance, and the game character, `PlayerAvatar` classes, for instance, as well as defining actor properties and functions in C++.

Chapter 6, Creating Game Actors, provides steps to write code and set up the main character for the *Pangaea* game. It includes setting up the character, creating the animation instance, defining the state machine, and synchronizing the animations.

Chapter 7, Controlling Characters, provides methods of controlling game characters. This includes configuring the input map, handling player input, and effectively processing the reactions of the player character. Additionally, you will be introduced to the AI controller and the navigation system for controlling non-player characters.

Chapter 8, *Handling Collisions*, discusses the engine's collision system and its configurations for game interactions. To handle collision events – attack hits and projectile hits, for example – you will learn how to configure actor colliders and triggers. Using ray casts to check whether a projectile hits the target is also introduced in this chapter.

Chapter 9, *Improving C++ Code Quality*, presents how to employ software engineering practices during code refactoring and refinement. This chapter implements class generalization, caching, and pooling methods to improve the game code's quality and performance.

Chapter 10, *Making Pangaea a Network Multiplayer Game*, starts by introducing the fundamental concepts related to multiplayer games, including servers, clients, and multiplayer modes. You will be guided step by step through converting the single-player *Pangaea* game into a multiplayer game.

Chapter 11, *Controlling the Game Flow*, intends to make *Pangaea* a complete multiplayer game, which has a main menu as the lobby, so that players can decide whether they want to start a host or join a game session. C++ and Blueprint scripting skills for user interface operations are also revealed in this chapter.

Chapter 12, *Polishing and Packaging the Game*, provides resources, methods, and suggestions on how to polish games from both visual experience and product quality aspects. This chapter also provides steps for configuring and packaging the *Pangaea* project to be an executable standalone game for distribution.

To get the most out of this book

You will need to have knowledge and experience in using Unreal Engine. Basic Blueprint scripting knowledge is also required prior to reading this book.

Software/hardware covered in the book	Operating system requirements
Unreal Engine 5.0 and up	Microsoft Windows 10 and up
Microsoft Visual Studio 2002 with the C++ compiler	

If you conduct experiments with the samples on systems other than Microsoft Windows, such as macOS, please keep in mind that there may be user interface and configuration differences that may not be addressed in this book.

If you are using the digital version of this book, we advise you to type the code yourself or access the code from the book's GitHub repository (a link is available in the next section). Doing so will help you avoid any potential errors related to the copying and pasting of code.

Download the example code files

You can download the example code files for this book from GitHub at `https://github.com/PacktPublishing/Unreal-Engine-5-Game-Development-with-C-Scripting`. If there's an update to the code, it will be updated in the GitHub repository.

We also have other code bundles from our rich catalog of books and videos available at https://github.com/PacktPublishing/. Check them out!

Conventions used

There are a number of text conventions used throughout this book.

Code in text: Indicates code words in text, database table names, folder names, filenames, file extensions, pathnames, dummy URLs, user input, and Twitter handles. Here is an example: "The AProjectile class can be inherited as child classes for creating various fireable objects, such as AFireBall, AMissile, ABomb, and so on."

A block of code is set as follows:

```
#pragma once
#include "CoreMinimal.h"
#include "GameFramework/Actor.h"
#include "DefenseTower.generated.h
```

When we wish to draw your attention to a particular part of a code block, the relevant lines or items are set in bold:

```
void APangaeaCharacter::BeginPlay()
{
...
_AnimInstance = Cast<UPangaeaAnimInstance>(
GetMesh()->GetAnimInstance());
...
}
```

Any command-line input or output is written as follows:

```
$ mkdir css
$ cd css
```

Bold: Indicates a new term, an important word, or words that you see onscreen. For instance, words in menus or dialog boxes appear in **bold**. Here is an example: "In the **Unreal Project Browser** window, choose the **GAMES** tab on the left side. Then select the **First Person** template."

> **Tips or important notes**
> Appear like this.

Additionally, the C++ sample code provided in this book adheres primarily to Unreal Engine's coding standard, ensuring consistency and minimizing confusion for readers. For detailed information, you can visit the official *Code Standard* website here: `https://docs.unrealengine.com/5.0/en-US/epic-cplusplus-coding-standard-for-unreal-engine/`.

Exceptions may occur when using compact expressions that are clear and easily understood, allowing the text to fit within the constraints of the page printing layout without compromising reader comprehension. For example, the following line of code follows the code standard by explicitly declaring the type of the assigned `GameInst` variable:

```
UPlayerAvatarAnimationInstance* GameInst = Cast
  <UPlayerAvatarAnimationInstance>(GetMesh()->GetAnimInstance())
```

The following modified version is used instead:

```
auto GameInst = Cast<UPlayerAvatarAnimationInstance>(
  GetMesh()->GetAnimInstance())
```

Get in touch

Feedback from our readers is always welcome.

General feedback: If you have questions about any aspect of this book, email us at `customercare@packtpub.com` and mention the book title in the subject of your message.

Errata: Although we have taken every care to ensure the accuracy of our content, mistakes do happen. If you have found a mistake in this book, we would be grateful if you would report this to us. Please visit `www.packtpub.com/support/errata` and fill in the form.

Piracy: If you come across any illegal copies of our works in any form on the internet, we would be grateful if you would provide us with the location address or website name. Please contact us at `copyright@packt.com` with a link to the material.

If you are interested in becoming an author: If there is a topic that you have expertise in and you are interested in either writing or contributing to a book, please visit `authors.packtpub.com`.

Share Your Thoughts

Once you've read *Unreal Engine 5 Game Development with C++ Scripting*, we'd love to hear your thoughts! Scan the QR code below to go straight to the Amazon review page for this book and share your feedback.

https://packt.link/r/1-804-61393-2

Your review is important to us and the tech community and will help us make sure we're delivering excellent quality content.

Download a free PDF copy of this book

Thanks for purchasing this book!

Do you like to read on the go but are unable to carry your print books everywhere?

Is your eBook purchase not compatible with the device of your choice?

Don't worry, now with every Packt book you get a DRM-free PDF version of that book at no cost.

Read anywhere, any place, on any device. Search, copy, and paste code from your favorite technical books directly into your application.

The perks don't stop there, you can get exclusive access to discounts, newsletters, and great free content in your inbox daily

Follow these simple steps to get the benefits:

1. Scan the QR code or visit the link below

https://packt.link/free-ebook/9781804613931

2. Submit your proof of purchase
3. That's it! We'll send your free PDF and other benefits to your email directly

Part 1 –
Getting Started with
Unreal C++ Scripting

In this part, the primary focus will be on providing an introduction to the basics of C++ programming, specifically for game development with Unreal Engine. You will gain knowledge on creating a C++ game project in Unreal, as well as utilizing MS Visual Studio to access and modify the game's source code. Moreover, essential concepts of C++ and object-oriented programming, along with their syntax, will be introduced. Building upon this foundation, we will examine the generated source code to conduct an initial investigation into the game project.

This part contains the following chapters:

- *Chapter 1, Creating Your First Unreal C++ Game*
- *Chapter 2, Editing C++ Code in Visual Studio*
- *Chapter 3, Learning C++ and Object-Oriented Programming*
- *Chapter 4, Investigating the Shooter Game's Generated Project and C++ Code*

1

Creating Your First Unreal C++ Game

Unreal Engine (UE) is one of the most popular 3D computer graphics game engines developed by Epic Games, providing a comprehensive set of tools and functionalities to develop high-quality, immersive 3D simulations. The engine offers its intuitive visual scripting system, **Blueprint**, and a robust **C++** programming framework for developers of all skill levels. This book provides a concise introduction to C++ programming and demonstrates how to write C++ scripts in UE for game development.

In this chapter, you will learn the essential skill of creating an Unreal C++ project from scratch or converting an existing Unreal Blueprint project into an Unreal C++ project, which serves as a fundamental skill to advance in game development. By mastering this process, you will gain the necessary foundation to take your game development abilities to the next level.

This chapter will cover the following topics:

- Understanding C++ scripting in Unreal
- Creating your C++ shooter project from a template
- Converting an existing Blueprint project to a C++ project

Technical requirements

As a reader of this book, you will be expected to have common computer operational skills. You should also have basic knowledge of and experience with the UE5 editor, as well as some Blueprint scripting skills.

To follow this chapter, you should have installed Epic Games Hub and the 5.03 or later version of the engine editor on your computer. If you haven't done so, please go to the official Epic website (https://www.unrealengine.com/en-US) to register an account and download the Epic Games Launcher.

The minimum required development environment is as follows:

- **Operating system**: Windows 10
- **Processor**: Intel 7th generation or equivalent
- **Memory**: 16 GB of RAM
- **GPU**: GTX 1080 (or AMD equivalent)
- **DirectX**: Version 12
- **Storage**: 25 GB of available space
- **Additional notes**: 8 GB of VRAM recommended

The official system requirements can be found here: `https://docs.unrealengine.com/5.0/en-US/hardware-and-software-specifications-for-unreal-engine/`. To save game editing time in the UE5 editor, it is recommended to use a computer with an i9 (or an AMD equivalent) CPU, 64 GB of RAM, and a GeForce RTX 3060 video card.

Understanding C++ scripting in Unreal

Before getting started, we need to answer some questions that people usually ask about **C++ scripting**. This will help to clarify the pros and cons of using C++, the reasons to use C++, and the difference between UE C++ scripting and C++ programming.

What is the difference between C++ and Blueprint?

Both C++ and Blueprint are scripting languages that can accomplish the same tasks, but one might be better suited than the other under certain circumstances. The main difference between them is that C++ is a programming language that allows you to write general-purpose, text-based code, whereas Blueprint is a visual scripting system for UE.

For UE projects, game studios usually use both C++ and Blueprint to develop commercial-level games. C++ is usually used for advanced techniques, complex algorithms, and big-scale logic code. If you can script with C++, you will have more chances to work on a professional team.

One of the most important advantages of using C++ is performance. C++ allows you to write low-level operational code. It also provides control over the core system that is not accessible to Blueprint. In addition, the final C++ code will eventually be optimized and compiled to be machine-friendly binary native code. On the other hand, Blueprint scripts are interpreted and executed by a middle layer, which means more execution time.

C++ code and files can be well-organized based on an entire project's mechanics. It is easy to globally search, locate, and access code blocks to edit, maintain, and troubleshoot. In the meantime, it is also easier to read and understand a big chunk of code that implements complex algorithms and logic. Blueprint, on the other hand, is a context-sensitive scripting environment. Blueprint graphs are relatively independent. When a graph needs to solve complex logic, the nodes and the connection lines create messy spaghetti that can hardly be understood and maintained.

C++ also has some shortcomings. One example is that it may cause critical errors that may crash an entire system. That is usually caused by the developer's mistakes. Since Blueprint is a protected layer, it is safer, and hence, the chances of the system crashing are fewer.

In conclusion, the choice between C++ and Blueprint should be made based on specific development requirements and conditions, considering the pros and cons of each approach.

When do you use C++?

Both C++ and Blueprint can handle game development processes without a problem. There is no exact rule that regulates when to use C++ or Blueprint. It mainly depends on your experience and the actual needs of different games. You make your own decision based on how much you know about the two scripting systems.

Before you start working on something, you can ask yourself this question: *"Where does it make sense to use C++, and where does it make sense to use Blueprints?"* We recommend basing your answer on the following aspects and trade-offs:

- Performance
- Logic and algorithm complexity
- Accessibility to a system's core functions
- The developer's experience

If you want higher performance and deal with advanced game logic and system processes, and you are capable of coding and solving complex problems, you should go for C++.

What is the difference between C++ programming and C++ scripting?

You may be confused about the difference between C++ programming and C++ scripting. We want to clarify the meanings of these two terms.

C++ programming means using the C++ programming language to write code for any purpose; it doesn't have to be just for UE projects. C++ scripting, in this book, is a specific dialect of the C++ programming language supported by the UE. It takes advantage of the power of C++ syntax and also works with UE's **Application Programming Interfaces** (**APIs**), which allow developers to create and extend the engine's functionalities for their games and the development environment's context, such as objects, graphics, audio, and network communication.

Now that we have a basic overview of C++ and have learned why and when to use C++ for Unreal game developments, let's dive deeper into C++ scripting by creating a sample project.

Creating your C++ Shooter project from a template

Now, it's the time to get your hands dirty working on a UE5 C++ project yourself. We will go through the steps to create a new C++ project from scratch based on the **First Person** template.

The **First Person** template is one of the default game templates that come with UE. When you want to create a new project, you can pick this template from the **Unreal Project Browser** window. Our new *MyShooter* game will derive all the features from the template game, and we don't have to do any additional work.

To get started with C++ scripting, we first need to install an IDE. In this book, we will use MS Visual Studio 2022 as an example.

Installing Visual Studio 2022

Visual Studio (**VS**) is an **Integrated Development Environment** (**IDE**) from Microsoft. It is a tool used to create, edit, debug, and compile code. In order to do C++ scripting, you need to go to the official website at `https://visualstudio.microsoft.com/vs/` and download the **Community 2022** version installation package (see *Figure 1.1*).

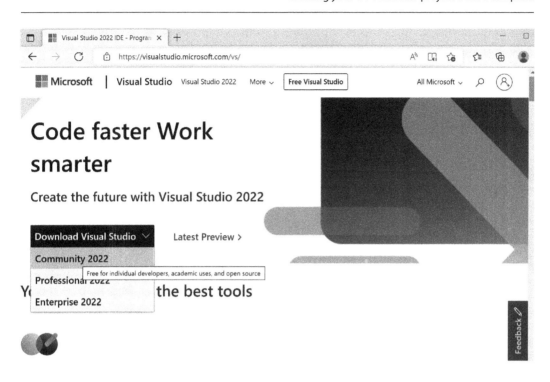

Figure 1.1 – Downloading VS 2022

> **Note**
>
> To install VS, a Microsoft account is typically required. If you don't have a Microsoft account, you can register using the following page: `https://account.microsoft.com/account/`.

Next, launch `VisualStudioSetup.exe` inside the folder where you downloaded the VS installer (the `\Downloads` folder, for example).

Enable the two **Game development with C++** and **Desktop development with C++** checkboxes – these two options tell the installer to install the C++ compiler and the professional game development support for UE (see *Figure 1.2*).

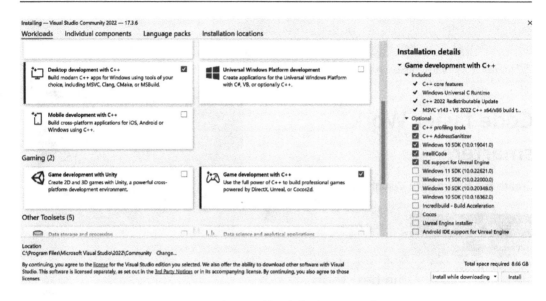

Figure 1.2 – Picking workloads for the VS installation

Also, keep an eye on the following options on the **Installation details** panel that belongs to the **Desktop development with C++** group, and make sure the following are checked:

- **C++ profiling tools**
- **C++ AddressSanitizer**
- **Windows 10 SDK**
- **IntelliCode**
- **IDE support for Unreal Engine**

Then, click the **Install** button to install the workloads and reboot the system, and then you will see a prompt from the dialog popup (see *Figure 1.3*):

Done installing

Visual Studio has been successfully installed. We recommend rebooting soon to clean up any remaining files.

OK

Figure 1.3 – The VS Done installing dialog box

The next thing we need to do is to confirm that we have installed the engine source code together with the UE5 editor. The reason why we need this is that when we generate a new project, the engine source code can be integrated into the new project; under certain circumstances, we may need to modify or customize the engine for the game's specific needs.

Ensuring your UE has the source code installed

Before launching the UE5 editor, we first need to check whether **Engine Source** is installed for the editor. By doing this check, we make sure that the UE5 source code is integrated with the C++ projects we are going to create.

The three steps to check or install the engine source code are as follows:

1. Click the downward arrow button and choose **Options** from the drop-down menu.

2. Make sure that the **Engine Source** option is checked.

3. Press the **Apply** button:

Figure 1.4 – The UE5 Options menu

UE is an ongoing development product, with bugs and defects that may need to be fixed by its users. Also, professional developers sometimes modify the engine source code to adapt to their specific needs. An example of this is when we face an issue with geometry instancing (or instanced rendering) working only in the game's development build but not in the release build, which is subsequently resolved by our engineer modifying the engine's source code.

> **Note**
>
> Geometry instancing is a rendering technique that renders multiple instances of a visual object in a single draw call and provides each instance with some unique attributes: `https://en.wikipedia.org/wiki/Geometry_instancing`.

We are now ready to start the UE editor through the Epic Games Launcher.

Launching the UE5 editor through the Epic Games Launcher

Launching the UE5 editor is pretty straightforward. You simply click the **Launch** button on the 5.03 engine card to start the editor (see *Figure 1.5*).

Figure 1.5 – Launching the UE5 editor from the Epic Games Launcher

The next thing we want to do is to create a new game project. Let's name the new project MyShooter.

Creating the MyShooter C++ project

To create the project, follow these steps (and see *Figure 1.6* for reference):

1. In the **Unreal Project Browser** window, choose the **GAMES** tab on the left side.
2. Select the **First Person** template.
3. Select the **C++** button.

4. Choose the project location (for example, C:\UEProjects) and type MyShooter in the **Project Name** field.

5. Click the **Create** button.

Figure 1.6 – Creating the MyShooter project

The created game project also includes the starter content, which is packaged with assets and resources that can be used to prototype the game.

The engine will do some initialization work and then open the editor when things are ready. If you look at the project tree panel's **MyShooter** tab in the bottom-left corner of the editor window, you should see the **C++ Classes** node on the same layer as the **Content** node (see *Figure 1.7*).

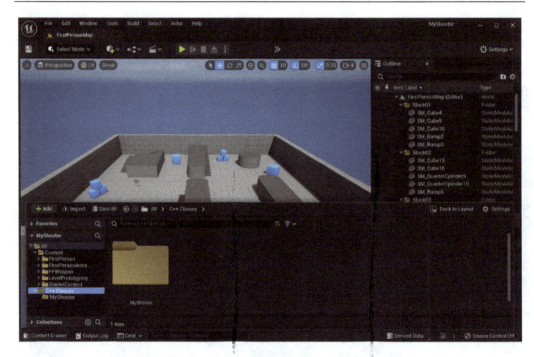

Figure 1.7 – The MyShooter C++ project opened in the UE5 editor

Associating VS with UE5 as the default source code editor

Since we created the C++, project, all the C++ source code for the game was already generated. To open the source files directly in the UE5 editor, we want to associate VS as the engine editor's default IDE.

On the UE5 Editor's main menu, select **Edit | Editor Preferences** to open the preference window, then find the **General | Source Code** item on the left panel, and finally, pick **Visual Studio 2022** from the **Source Code Editor** dropdown (see *Figure 1.8*).

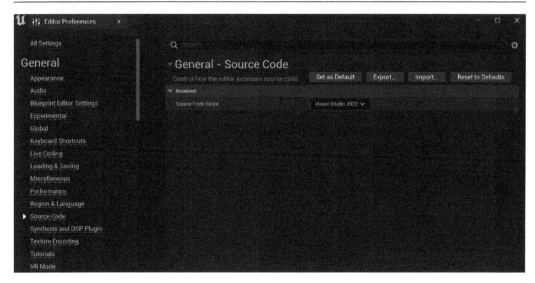

Figure 1.8 – Making VS the default source code editor

You can now use VS to open the source code files.

Opening the C++ source code in VS (optional)

If you want to open and view the C++ source code in VS, you can find the source code file (for example, C++/MyShooter/MyShooterCharacter.cpp) in the project and simply double-click on it (see *Figure 1.9*).

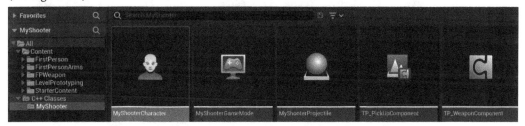

Figure 1.9 – Opening MyShooterCharacter.cpp source code in VS

The system will automatically launch VS, and the VS editor will open the MyShooterCharacter.cpp file (see *Figure 1.10*).

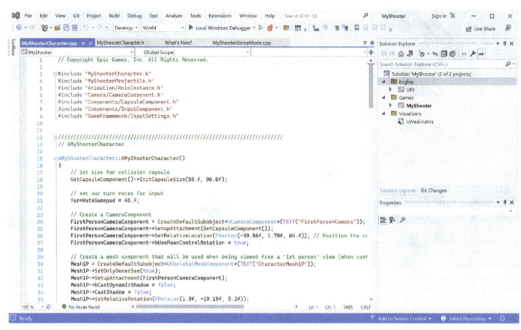

Figure 1.10 – Viewing the MyShooterCharacter.cpp source code in VS

Back in the Unreal editor, click the **Play** (▶) button to start the game. While playing the game on the battlefield, you can control your character, move them around, and pick up the gun in front of them (see *Figure 1.11*).

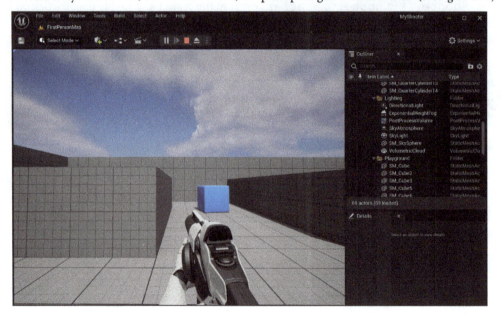

Figure 1.11 – Playing the MyShooter game

We have learned how to create a **UE** C++ project from scratch. However, what if we already have a Blueprint project and want to convert it to a C++ project? UE allows developers to do it by adding a new C++ class to the project. Let's practice converting a *MyBPShooter* Blueprint project.

Converting an existing Blueprint project to a C++ project

UE provides a very straightforward way to convert an existing Blueprint project to a C++ project. All you need to do is add a C++ class to your project and then let UE take care of the conversion and add the needed project files:

1. First of all, you have to create a Blueprint project, *MyBPShoopter*, under C:\UEProjects (you can choose a different path to create the new project). Use the same steps introduced in the *Creating the MyShooter C++ project* section, but choose **BLUEPRINT** instead of **C++** for the creation of the *MyBPShooter* project.

Figure 1.12 – Creating MyBPShooter in UE5

2. Secondly, open the new project in UE5. Pay attention to the project tree; it doesn't have the **C++ Classes** node at this stage.

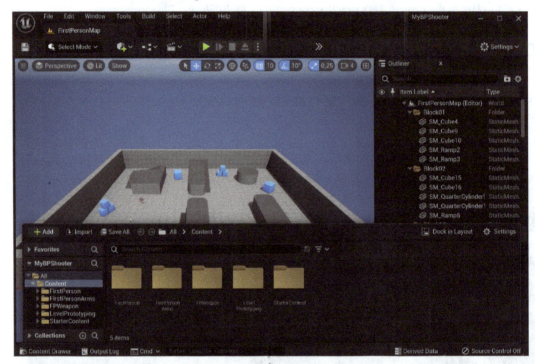

Figure 1.13 – Open MyBPShooter in UE5

3. Select **Tools | New C++ Class** from the editor's main menu, and then, in the **Add C++ Class** window (see *Figure 1.14*), choose **Character** as the base class (a class that contains common attributes and methods that are shared by its derived classes) to create the MyShooterCharacter class.

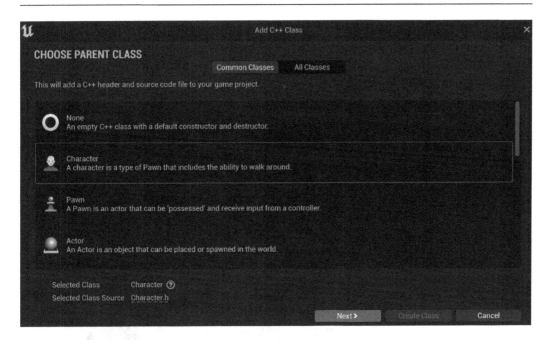

Figure 1.14 – Adding a new C++ class from the Character class

Once you click the **Next>** button, it will navigate to the **NAME YOUR NEW CHARACTER** screen.

4. On the **NAME YOUR NEW CHARACTER** screen, type MyBPShooterCharacter into the **Name** field.

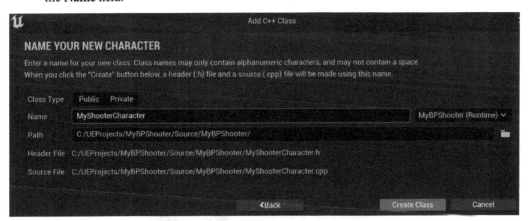

Figure 1.15 – Adding the MyBPShooterCharacter C++ class

Please pay attention to the path where the header and the source files will be placed. They look different from the *MyShooter* project because the C++ node hasn't been created yet. Don't worry about it at the moment. Once the conversion job is done, the system will automatically move the files to the right place.

5.   After clicking the **Create Class** button, you will see a progress bar.

Figure 1.16 – The MyBPShooterCharacter C++ class Adding code to project… progress bar

Wait for the pop-up message, which indicates that the C++ class job has been added.

Figure 1.17 – A message saying that the MyBPShooterCharacter C++ class is now added

6.   Click the **OK** button. Now, you will see the message dialog, which asks you whether you want to edit the code (see *Figure 1.18*). Choose **No** here.

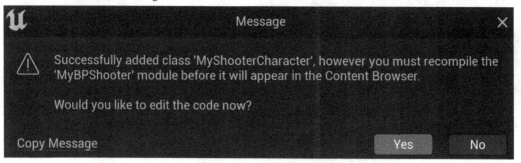

Figure 1.18 – Dialog for editing the MyBPShooterCharacter source code

7.   Shut down your UE editor and reopen MyBPShooter. When you see a dialog that asks whether you want to rebuild the project, answer **Yes** here.

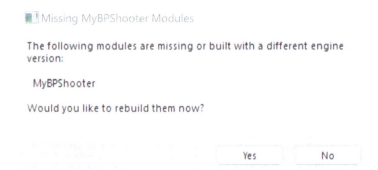

Figure 1.19 – The rebuilding MyBPShooter dialog

When it is done, you will find the new **C++ Classes** node on the project tree, and the `MyShooterCharacter` class is already placed in the `MyBPShooter` folder:

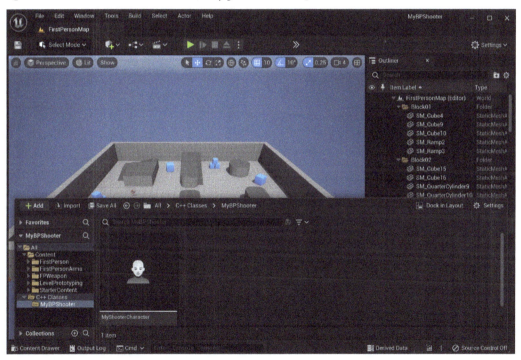

Figure 1.20 – The converted MyBPShooter C++ project

You may have noticed that some other files, such as `MyBPShooterGameMode` are missing, in comparison with the *MyShooter* project. That is because the Blueprint versions already exist, so the corresponding C++ versions are not automatically generated. You can choose to manually convert those blueprints to C++ classes only when necessary; otherwise, you just keep the blueprints.

Summary

In this chapter, we introduced C++ and the advantages of using it for professional game development. Then, you practiced creating the new *MyShooter* C++ project and converting the *MyBPShooter* Blueprint project to a C++ project. Plus, you also set up the development environment with VS and the C++ solution files.

In the next chapter, we will first walk through each part of the IDE's user interface. Then, you will create a C++ project and practice writing some simple C++ code. Some code editing tricks will be introduced while editing your code.

2
Editing C++ Code in Visual Studio

Are you new to coding in general? Then you need to use an editing tool!

C++ source code is just regular text files named with some special extension names, such as `.cpp`, `.h`, and so on. You can basically use Windows **Notepad** to open and edit C++ source code files. However, since Notepad is a basic editing tool that lacks functionalities, we recommend using **Visual Studio** (**VS**) as the code editor.

Why use VS? VS is a feature-rich **integrated development environment** (**IDE**) that supports many aspects of software development. It empowers you to complete the entire development cycle in one place. You can use VS to create, edit, debug, test, and build your code. VS also has the most popular programming language compilers integrated with the installation package so that C++ source code can be directly compiled to be executable machinery code. Moreover, VS especially supports Unreal Engine and works well with the engine's development environment.

By following the step-by-step journey of this chapter, you will get to know the IDE's **user interface** (**UI**), be capable of creating and writing C++ code, and learn how to build C++ solutions to generate standalone executables. This chapter includes the following sections:

- Launching VS
- Walking through the VS IDE's UI
- Editing code in VS
- Practicing C++ coding

Technical requirements

To explore the creation of C++ projects and editing C++ code, it is necessary to have VS installed on your system.

VS has both Windows and macOS versions. It also has Community, Professional, and Enterprise editions. The examples of this book are based on the VS 2022 Windows Community edition.

Since VS is an IDE that you will use for C++ scripting, being familiar with the development environment and the scripting skills is a prerequisite.

The code for this chapter can be found at `https://github.com/PacktPublishing/Unreal-Engine-5-Game-Development-with-C-Scripting/tree/main/Chapter02/MyCPP_01`.

Launching VS

In *Chapter 1*, we went through the installation of VS, so you should already have installed VS on your system. Since VS is an independent application, you can launch it either from the **operating system (OS)** or in Unreal Engine.

In Windows, simply search for `virtual studio` and pick the version of the IDE that you wish to launch:

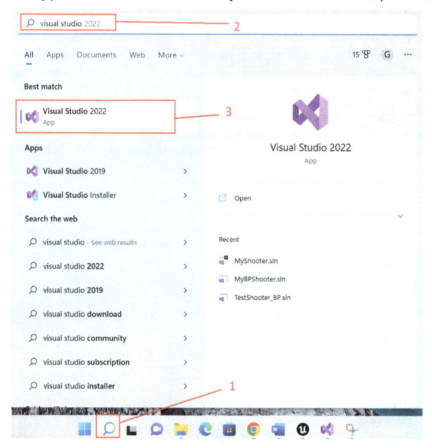

Figure 2.1 – Starting VS in Windows

Now, let's practice launching VS in Unreal Engine. Say we want to open the `MyShooterCharacter.cpp` file—you first need to find `MyShooter/All/C++ Classes/MyShooter` on the **Content Drawer**, and then you can double-click on the **MyShooterCharacter C++ Class** item:

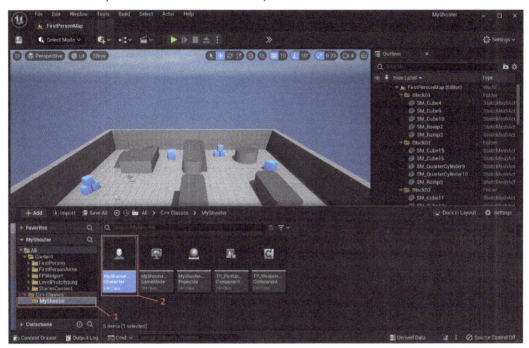

Figure 2.2 – Starting VS in Unreal Engine

This operation will launch VS if it hasn't been launched yet and open the `MyShooterCharacter.cpp` file in the editor:

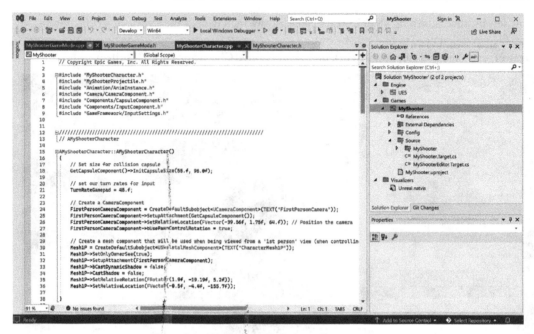

Figure 2.3 – VS opened MyShooterCharacter.cpp

Now, you should have your Unreal game development environment installed and set up. The engine editor and VS are both open. Next, we'll take a close look at the IDE's UI.

Walking through the VS IDE's UI

VS is a powerful and complex tool set. This book only covers features and functionalities that you will need for learning C++ scripting. You can visit Microsoft's official websites to learn more about VS in the future. Here are the links:

- VS IDE documentation: `https://learn.microsoft.com/en-us/visualstudio/ide/?view=vs-2022`

- Learn to code in VS: `https://visualstudio.microsoft.com/vs/getting-started/`

Once you have opened VS, you will see the IDE editor:

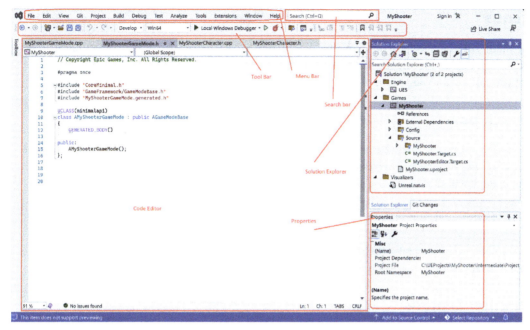

Figure 2.4 – VS IDE editor

Using *Figure 2.4*, let's take a look at all the important elements present in the editor.

Code editor

The code editor on the left side is the place where you write your C++ code. You can open multiple source code files in the code editor. The opened filename tabs are displayed at the top of the editor. *Figure 2.5* shows four source code files opened in the editor: `MyShooterGameMode.cpp`, `MyShooterGameMode.h`, `MyShooterCharacter.cpp`, and `MyShooterCharacter.h`:

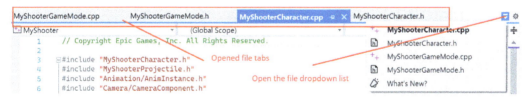

Figure 2.5 – VS code editor open file tabs

You can select to view any source file by clicking on the respective tab. In the case where too many files are opened, and not all the file tabs can be shown on the tab bar, you can click the drop-down button in the top-right corner to show a list of all open files.

Menus

The menu bar along the top of VS groups commands and categories. Each menu item has its own drop-down vertical sub-menus. The menus provide interfaces to the operations, options, and features that you can use for your coding work.

Search box

The search box on the menu bar is a special tool that helps you to find IDE menus and options, while also searching your code. This feature offers a quick and easy way to search across IDE features and your code. It will show you a list of options that are relevant to the entered text.

Toolbar

The toolbar is a horizontal strip beneath the menu bar that contains some shortcut buttons that are bound to commands, such as **Open File**, **Save**, **Save All**, **Start**, **Start Debug**, and so on.

Solution Explorer

Solution Explorer shows a solution's projects, folders, and files in a hierarchical tree representation. You can browse the tree to select and open files in the editor.

A VS solution file is a file that organizes multiple projects into a single solution. It assists developers in managing the various files and dependencies within a project, making complex software applications easier to work on and build. The solution file contains information about projects, their configurations, and their dependencies, which allows VS to compile and build the entire solution.

A VS solution may contain multiple projects, and each project has its own project, source code, configuration, dependencies, and other files. In this case, the MyShooter solution contains two projects, **UE5** and **MyShooter**:

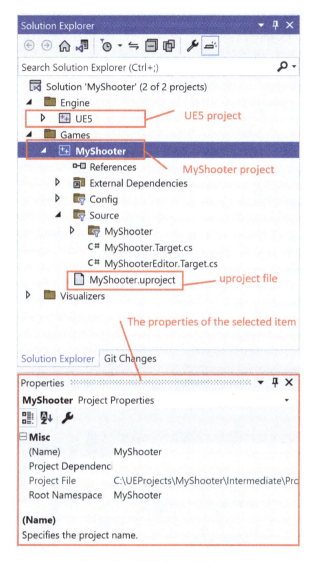

Figure 2.6 – VS Solution Explorer

In *Figure 2.6*, the **Properties** window shows the properties of the selected item—in this case, the properties of **MyShooter**. The **MyShooter** project contains the `Myshooter.uproject` Unreal project file.

Output window

The **Output** window shows output messages from the code-building process. Let's build the project to see the build outputs.

Choose **Build Solution** from the **Build** menu. The **Output** window should then obtain the screen focus and show the build result (see *Figure 2.7*). You can also manually open the **Output** window by going to **View | Output**:

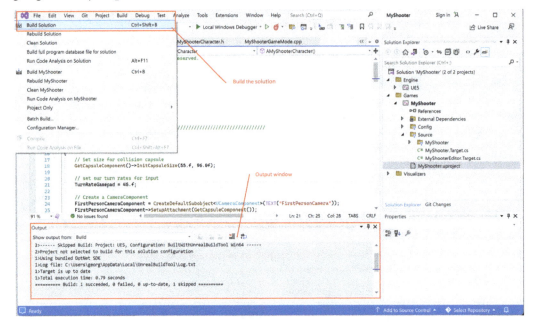

Figure 2.7 – VS Build and the Output window

In *Figure 2.7*, the **Output** window informed us that the building process succeeded without any errors.

Error List window

The **Error List** window shows build errors, warnings, and messages about the current state of your code. When building your code, if the code has errors, warnings, or messages, the **Error List** window is automatically opened. You can also manually open the **Error List** window by going to **View | Error List**.

As an example, let's make a small change to the `MyShooterCharacter.cpp` code by deleting the semicolon (;) at the end of *line 18*:

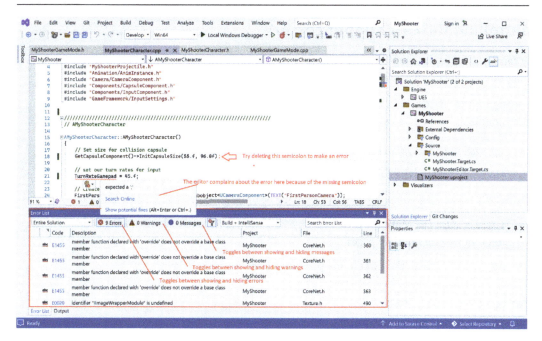

Figure 2.8 – VS Build and the Error List window

This code error will automatically cause the editor to complain by showing the `TurnRateGamepad` variable name with a red squiggly underline. From the **Build** menu, choose **Build Solution**, and you will see that our change caused nine errors.

The **Errors**, **Warnings**, and **Messages** toggle buttons can enable or disable filters so that you can choose to view information you are interested in.

Please remember to restore the deleted semicolon after the test. Reopen the **MyShooter** project in UE5, and answer **Yes** to rebuild `MyShooter`. Then, open the C++ solution in VS, and then choose **Build Solution**. This will clear the generated errors.

You should now have a basic idea about the VS UI. It is time to learn how to edit code in the IDE.

Editing code in VS

The VS IDE has ample powerful editing features and tools that can help developers effectively create and edit their code. Almost all the tools can be found on the main menu system. Here, we'll only introduce the commonly used editing tools and the shortcut keys that can help you to get started. It is recommended to utilize and practice the following shortcuts and keys in your later exercises.

Controlling the caret (input cursor)

The caret is often represented as a blinking vertical line when editing in VS. It indicates the current input point and determines where new text will be entered.

Knowing how to use the keys for caret navigation is an essential editing skill, so let's find out a bit more about this:

- *Up*, *Down*, *Left*, *Right arrow keys*: Move the caret up, down, left, or right, respectively
- *Home*, *End*: Move the caret to the beginning or the end of a code line, respectively
- *Ctrl + Home*, *Ctrl + End*: Move the caret to the beginning or the end of the code file, respectively
- *Page Up*, *Page Dn*: Move the caret one page up or down, respectively, when a big code file needs to be viewed on multiple pages
- *Mouse click*: Moves the caret to the position where you just clicked

The text editing keys

The text editing keys are used to toggle between **Insert** and **Overwrite** modes, as well as to facilitate text deletions. Here's what they do:

- *Insert* (*Ins*): Toggles between the **Insert** and **Overwrite** modes.
- *Delete* (*Del*): Deletes the character on which the caret is. When a block of text code is selected, it deletes the entire selected code.
- *Backspace*: Deletes the character before the caret's current position and moves the caret to the deleted character's position.

Code selection

Code selection allows developers to highlight specific portions of text for actions such as copying, cutting, deleting, and so on. Code blocks can be selected not only by clicking and dragging with your mouse but also using these key combinations:

- *Shift + Left arrow key*, *Shift + Right arrow key*: Select one more character on the left or right side, respectively, when extending a selection area. Deselect a character on the left or right side, respectively, when shrinking a selection area.
- *Shift + Up arrow key*, *Shift + Down arrow key*: Select up or down one more line, respectively, when extending a selection area. Deselect up or down one line, respectively, when shrinking a selection area.
- *Ctrl + Shift + Left arrow key*, *Ctrl + Shift + Right arrow key*: Select one more meaningful section—such as a word, a variable, a symbol, and so on—on the left or right side, respectively, when extending a selection area. Deselect a meaningful section on the left or right side, respectively, when shrinking a selection area.

IntelliSense

IntelliSense is a valuable resource when you are writing code. It is like a smart agent working in the background that can show you relevant information about an object's name, available members of a type, or parameter details for different overloads of a method. IntelliSense can also be used to complete a word after you have typed in a sufficient number of characters to disambiguate it.

As with other text editing tools, VS supports basic editing operations, such as copy/paste and find/replace. Knowing the useful editing operations and the hotkeys will improve your coding performance.

Useful editing hotkeys

To enhance editing performance, as with most text editors, VS adheres to common editing conventions by supporting editing hotkeys.

Copy and paste

Use the following keys to copy and paste text:

- *Ctrl* + *C*: Copies the selected text
- *Ctrl* + *X*: Cuts the selected text
- *Ctrl* + *V*: Pastes the text on the clipboard at the current caret position

Find and replace

Use the following keys to search and replace texts:

- *Ctrl* + *F*: Searches for the keyword in the current document. Relocates the caret at the first found position.
- *Ctrl* + *Shift* + *F*: Globally searches for the keyword—`calculator`, for example—in all the solution files. A "matches found" list is displayed in the **Find calculator** window. This is a very useful tool when you are working on big-scale projects.
- *Ctrl* + *H*: Allows searching and replacing a keyword with different text inside the current document.
- *Ctrl* + *Shift* + *H*: Allows searching and replacing a keyword with different text for all source files in the solution.

Code block operations

You can comment and uncomment code by typing comment symbols, but these hotkeys will make it easier and faster to do this kind of work:

- *Ctrl* + *K* + *C*: Comments out the selected block of code
- *Ctrl* + *K* + *U*: Uncomments the selected block of code

- *Ctrl + K + F*: Automatically arranges the selected block of code to be the formatting settings
- *Ctrl + K + D*: Automatically arranges the document code to be the formatting settings

Go to operations

The **Go to** operation hotkeys can help you quickly navigate and locate the code you want to view. This is what they do:

- *F12, Ctrl + F12*: Goes to the definition or the declaration of the selected keyword, such as a variable, a class name, a function name, and so on.
- *Ctrl + G*: Goes to a line by a line number.
- *Ctrl + T*: Goes to all. This is a useful tool when you know a filename and want to open the file, especially when you're working on a big-scale project with hundreds or thousands of source files.
- *Ctrl + Tab*: Navigates between the most recent two document windows.

Debugging

Debug tools are used for troubleshooting errors and mistakes in your code, allowing you to pause the execution at certain breakpoints and do investigations. Debugging is a basic skill that programmers should have. The following are the basic debugging functions you need to know:

- *F9*: Toggles a breakpoint on a line of code. When running under the **Debug** mode, the program execution will pause at the breakpoint.
- *F5*: Starts running the program under the **Debug** mode.
- *Ctrl + F5*: Starts running the program without the **Debug** mode. Breakpoints will be ignored.
- *F10*: When pressed, the IDE will show you what it's doing one step at a time so that you can figure out where the problem is.
- *F11*: When pressed, the IDE will set the tracing focus on the function's first line so that you can dig deeper into that function for more details.

Now, it's time to apply the aforementioned editing skills to write code in a real C++ project.

Practicing C++ coding

Now, it's time to practice writing C++ code in VS. To simplify the learning process and avoid noise from the Unreal Engine code, we will use a pure C++ solution as the learning example.

Here are some recommendations to consider while editing the code:

- You don't need to fully comprehend the source code at this stage; simply copy the provided C++ code in this section and concentrate on the editor features, as the C++ programming syntax will be introduced in the next chapter.

- Try using the introduced VS editing keys as much as possible.

- Type the code manually instead of relying on copy and paste, as this will assist you in quickly mastering editing skills and becoming familiar with the editing environment.

So, let's get started.

Creating a new C++ solution in VS

Begin by starting VS from Windows and selecting **Create a new project**:

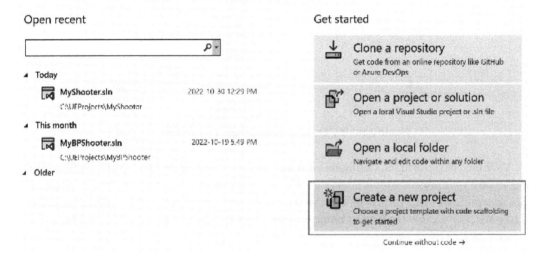

Figure 2.9 – VS: Creating a new project

Then, choose **Empty Project** on the **Create a new project** screen and click **Next**:

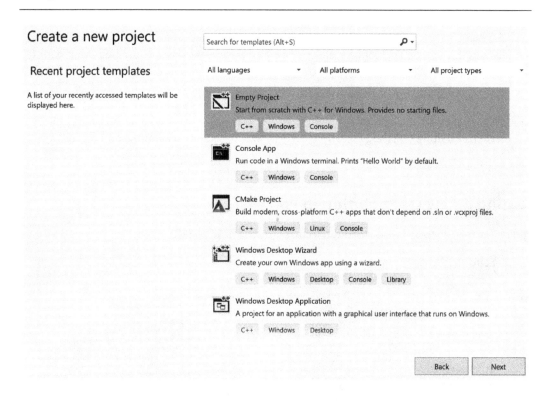

Figure 2.10 – VS: Creating a new empty C++ project

Now, VS should navigate to the **Configure your new project** screen. Here, you can choose the target folder to save your project. In this book, we are saving our examples into the `C:\C++Projects` folder, so create a `C++Projects` folder on your `C:` drive, and then select it in the **Location** box. You can type `MyCPP_01` into the **Project name** box and select `C:\C++Projects\` for the **Location** box (see *Figure 2.11*):

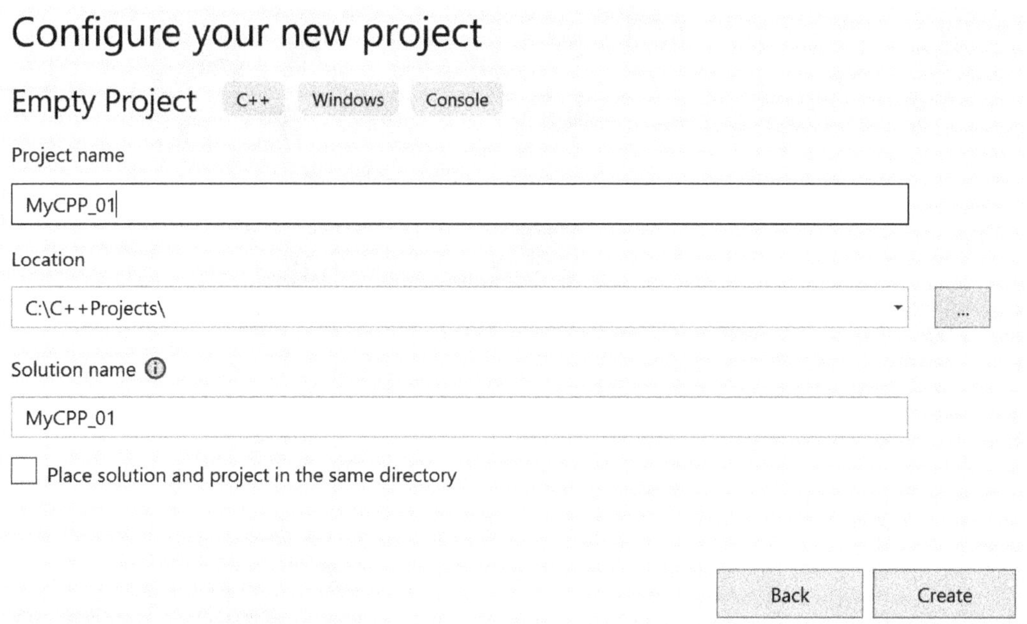

Configure your new project

Empty Project   C++   Windows   Console

Project name

```
MyCPP_01|
```

Location

```
C:\C++Projects\                                                    ▾        ...
```

Solution name ⓘ

```
MyCPP_01
```

☐ Place solution and project in the same directory

Back   Create

Figure 2.11 – VS: Configuring your new project

When it's done, click **Create** to proceed with the creation of the VS project solution.

When VS is launched, a `MyCPP_01.sln` solution file is already created and placed under the `C:\C++Projects\MyCPP_01` folder, and a `MyCPP_01.vxproj` project file is placed under the `C:\C++Projects\MyCPP_01\MyCPP_01` folder.

We can now add a C++ source code file to the project.

Creating the main.cpp file

In **Solution Explorer**, right-click on **Source Files**, then select **Add | New Item…**:

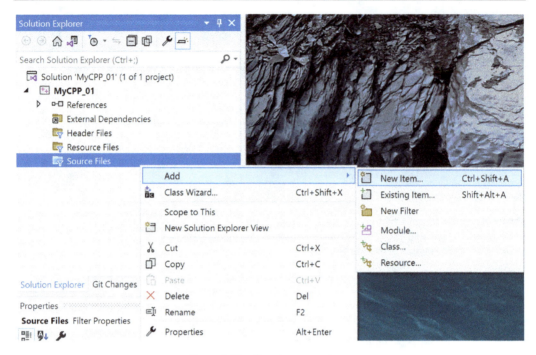

Figure 2.12 – Creating a new item

On the **Add New Item** window, choose **C++ File (.cpp)** and type `main.cpp` into the **Name** field. Click the **Add** button to create a `main.cpp` file:

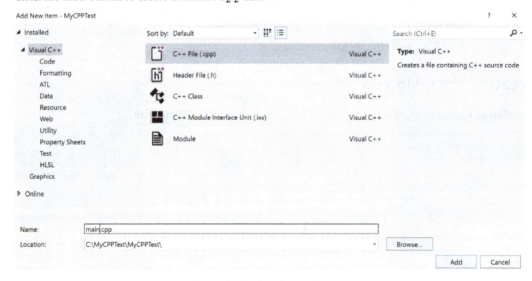

Figure 2.13 – Creating main.cpp

The VS editor should now have a **main.cpp** tab with the file open, and an input cursor in the editing area. In the **Solution Explorer** window, you will also see that the main.cpp file has been added as a child node under **Source Files**:

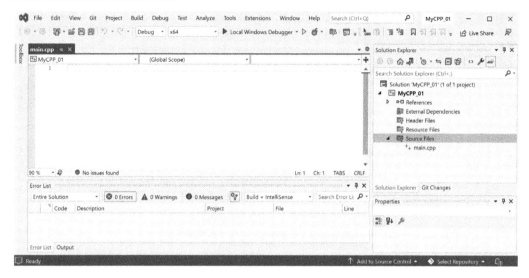

Figure 2.14 – main.cpp is created and ready for editing

We just worked on a C++ project in the VS IDE. By default, the IDE editor is set to use the **Light** color theme, but you can always customize it to your preferred color theme.

Changing the editor theme

You may have noticed that the screenshots of the Visual Studio editor captured for this book have black text on a white background. We purposely changed to the **Light** color theme for the IDE so that you can easily read the screenshots. However, you can choose to use your favorite color theme; for example, the **Dark** color theme may help reduce the chance of eye fatigue. To change the IDE's color theme, you can perform the following steps:

1. Open the menu bar and select **Tools** | **Options**.

2. In the **Options** list, select **Environment** | **General**.

3. In the **Color Theme** list, choose the color theme that you want to set—here, I have picked **Light**:

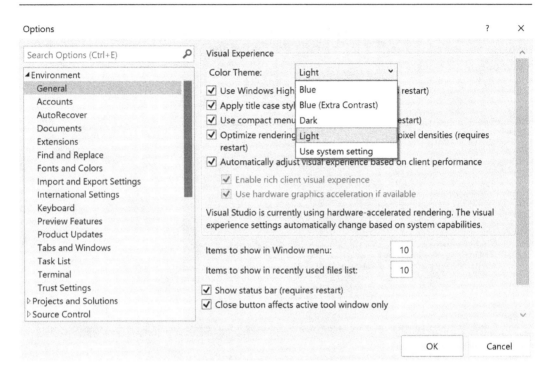

Figure 2.15 – Changing the color theme for the IDE

Now that we have created the `main.cpp` source code file, let's start writing some code to fill it up.

Writing the initial code for main.cpp

Now, we want to write the C++ code for `main.cpp`. C++ defines that the `main()` function is the starting point of the program. You can type or copy the following code into the editing area (you don't need to understand the source code at this moment; C++ programming will be explained in the next chapter):

```cpp
#include <iostream>

int main()
{
    std::cout << "MyCPP_01: Hello world!";
    return 0;
}
```

You can see the code in the editing area in *Figure 2.16*:

main.cpp

MyCPP_01   (Global Scope)   main()

```cpp
1    #include <iostream>
2
3    int main()
4    {
5        std::cout << "MyCPP_01: Hello world!";
6        return 0;
7    }
8
```

Figure 2.16 – Writing the code for main.cpp

Build the solution and play the program by clicking either the **Start** (▶) or **Start without Debug** (▷) button.

VS opens the **Debug Console** (a window in the VS editor that displays warnings, error messages, logs, and other useful information generated during execution time) and shows the output result, MyCPP_01: Hello world!:

Figure 2.17 – Running and outputting a message

We just added the main() entry function to the main.cpp module. Let's now add two more source files (Calculator.cpp and Calculator.h) to the project, which will contain the Calculator class's definition and implementation code.

C++ allows developers to create multiple source code modules. The benefits of using multiple code modules are set out here:

- Code is logically organized and grouped

- Module sizes are controllable

- It's easy to read and maintain the source code

- It's easy to search and locate code

So, let's get into it.

Adding the Calculator class

The `Calculator` class is designed to create a calculator object capable of performing addition and subtraction operations. To include the `Calculator` class in our project, we will add its header and source files to the project, as follows:

1. Add `Calculator.h` under `/MyCPP_01/Header Files` in **Solution Explorer**. Then, type in the following code:

    ```
    #pragma once

    #include <iostream>

    class Calculator
    {
    public :
    float Add(float a, float b);
    float Subtract(float a, float b);
    private:
    void OutputResult(float, std::string, float, float);
    };
    ```

2. Add `Calculator.cpp` under `/MyCPP_01/Source Files` in **Solution Explorer**. Then, type in the following code:

    ```
    #include "Calculator.h"

    float Calculator::Add(float a, float b)
    {
      float result = a + b;
      OutputResult(a, " + ", b, result);
      return result;
    }

    float Calculator::Subtract(float a, float b)
    {
      float result = a - b;
      OutputResult(a, " - ", b, result);
      return result;
    }

    void Calculator::OutputResult(float a, std::string op, float b,
    float result)
    ```

```
{
  std::cout << "Calculator: "
<< a << op << b << " = " << result << "\n";
}
```

If you inspect **Solution Explorer** now, `Calculator.h` and `Calculator.cpp` files should be included in the `MyCPP_01` project:

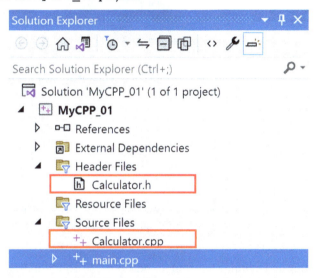

Figure 2.18 – Adding Calculator.h and Calculator.cpp

3. To test the calculator, change the `main.cpp` code to do some addition and subtraction calculations:

```cpp
#include <iostream>
#include "Calculator.h"

int main()
{
  std::cout << "MyCPP_01: Hello world! \n";

  Calculator Calculator;

  Calculator.Add(1.0f, 2.0f);
  Calculator.Subtract(10.0f, 5.0f);

  return 0;
}
```

4.   Now, run the program to see the output result:

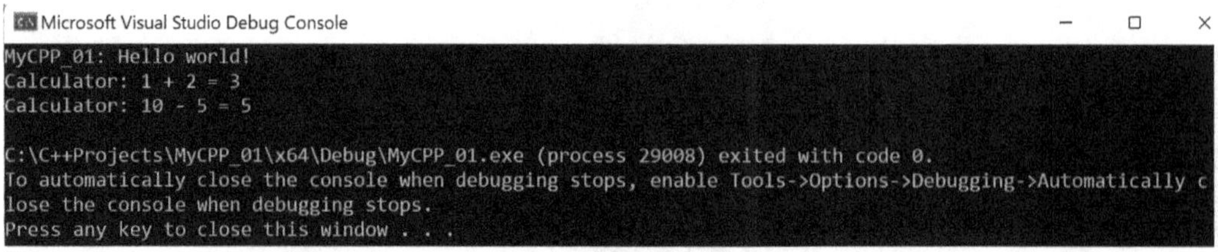

Figure 2.19 – MyCPP_01 output

Congratulations—you have now successfully developed a calculator application using C++!

Summary

By walking through the content in this chapter, you should have mastered basic code editing skills in VS. Being familiar with the IDE and the editing tools is fundamental for learning C++ scripting in the next chapters, and will also help in your future Unreal Engine game development practices.

The shortcut keys and the functions introduced in this chapter are particularly useful tools that you will want to remember and use when you edit your code—they will benefit you in terms of both your coding performance and code quality.

We also practiced creating and editing C++ source code in a C++ solution. Based on this, we will continue learning more C++ programming syntax, structural programming, and **object-oriented programing (OOP)** in the next chapter.

3

Learning C++ and Object-Oriented Programming

To use C++ to program games in Unreal Engine, you need to learn the C++ programming language. Almost all game engines need to support at least one scripting programing language because scripting provides interfaces that developers can use to control the game flow, integrate complex game logic, manipulate interactions between players and game entities, as well as process game events.

In order to utilize the Unreal Engine C++ APIs, it is essential to have a solid understanding of **Object-Oriented Programming** (**OOP**) principles and possess basic skills in C++ programming, which is what we will focus on in this chapter.

In this chapter, we will cover the following topics:

- What is C++?
- Exploring the C++ program structure
- Defining C++ functions
- Working with a basic calculator program
- Learning the C++ syntax
- Working on the improved calculator program
- Creating references and pointers
- Understanding OOP
- Working on an OOP calculator program

Technical requirements

The code for this chapter can be found at `https://github.com/PacktPublishing/Unreal-Engine-5-Game-Development-with-C-Scripting/tree/main/Chapter03`.

What is C++?

C++ is a programming language that supports OOP, can be used to create high-performance applications, and gives developers low-level control over system resources and memory. C++ is the most suitable programming language for certain applications that have special demands on performance and low-level system controls (for example, operating systems, embedded systems, game applications, and graphics rendering).

C++ is designed to be a compiled language, which means it is generally translated into machine language code that can be directly understood and executed by a system, making the generated code highly efficient:

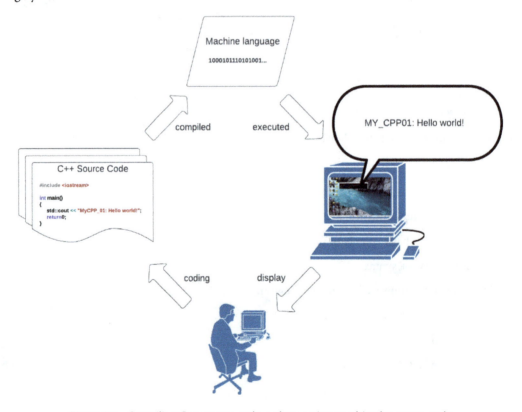

Figure 3.1 – Compiling C++ source code and executing machine language code

While writing C++ scripts in Unreal Engine, whenever you make changes, you need to compile your code first and then launch your game. Since Visual Studio comes with the C++ compiler, you should have no problem building your code in the IDE.

C++ is still evolving to adapt to modern programming trends. The language was first implemented in 1979, following which standard C++ versions were founded and implemented in the early 1990s. Versions after 2011 are named in the following format – the C++ prefix followed by a two-digit version number. C++20 is currently the latest version.

> **Note**
>
> C++ was developed as an extension of the C programming language, and both languages have almost the same syntax. The main difference between C++ and C is that the former supports OOP, which means when you learn C++, you also learn C.

Now that we have a basic idea about C++, let's start learning the fundamentals of the C++ programming syntax.

Exploring the C++ program structure

In C++, programs execute code line by line, with each statement typically terminated by a semicolon. A collection of code lines that perform specific tasks can be grouped as a function, enclosed by a pair of curly braces, with the function having a name followed by a set of parentheses.

For example, the main.cpp file we created in *Chapter 2* has two statement lines of code – the two lines of code are enclosed within a pair of curly braces, and the grouped block of code's function name is main (see *Figure 3.2*).

Figure 3.2 – The main.cpp code sample

C++ source programs generally follow the same program structure:

- #include statements at the beginning of the program, which allow this program to access the C++ system library and other C++ source program functionalities. #include statements are special statements that don't end with a semicolon.

> **Note**
>
> The C++ system library is a collection of pre-compiled functions and data structures that can be used by C++ programs. It provides a way to reuse code and avoid duplication of effort by allowing developers to access pre-written code instead of writing their own from scratch.

- The `Main()` function, which is run when the program is executed.

- `Return 0`, located at the end of the `main()` function, which notifies the completion result of running the program execution. Returning 0 usually means *success*, whereas some other values indicate error codes.

Let's go through the lines of code of `main.cpp`:

- *Line 1*: This contains the `#include` statement, which allows you to use the `std::cout` object and the `<<` operator (a symbol or keyword that is used to perform a specific operation on one or more operands), defined in the `iostream` library in the `main.cpp` file module

> **Note**
>
> The `iostream` header name is enclosed within two angle brackets (`< >`), which indicate that the header file is a system or library header file.

- *Line 3*: This defines the `main()` function

- *Lines 5 and 6*: These are the two statement lines terminated with semicolons

- *Lines 4 and 7*: This is a pair of braces that enclose the function body code

Having familiarized yourself with the basic structure of a C++ program, let's proceed to define some C++ functions.

Defining C++ functions

A **function** is a block of code that only runs when it is called. Functions are usually defined and used to perform certain actions, and they can be called anytime when needed; therefore, they are reusable code. Through the use of functions, we can avoid redundant code, reducing the risk of code inconsistency and the chance of program bugs.

Defining functions with or without parameters

To define a function, you should specify the name and return type of the function, followed by a pair of parentheses. Presented here is a function that has no parameter and a void return type:

```
void DisplayLabel()
{
   std::cout  << "The result of 1 + 2 is ";
}
```

You can pass data as parameters into a function, and the function parameters are placed in between the parentheses. Presented here is a function that has two int parameters and returns the addition of the two input values as the result:

```
int Add(int a, int b)
{
   return a + b;
}
```

Calling functions

To call a function, you can write the function's name followed by two parentheses. If the function requires a certain number of parameters, you should write the parameters so that the values will be passed into the function. Remember to put a semicolon at the end.

The following code snippet illustrates how to call the two functions defined earlier:

```
std:count << DisplayLabel() << Add(1, 2) << std:endl;
```

The output of the aforementioned line of code is the following:

```
The result of 1 + 2 is 3
```

Writing the main() function

Every C++ program must have a main function. It is a special function that indicates the start execution point of the program. A C++ program only has one main() function.

The return type of the main() function can be int or void, such as the following:

```
int main()
```

Or it can be the following:

```
void main()
```

The difference is that int main() tells the compiler that the program will return an integer value to the operating system, whereas void main() tells the compiler that the program will not return any value.

In order to better understand the basic C++ program structure, let's use the steps provided in the previous chapter to do an exercise by creating a MyCPP_02 project, which will do some simple calculations.

Working with a basic calculator program

The MYCPP_02 program should have a main() function and an Add() function. The signature and the tasks defined for these two functions are as follows:

- void Main(): This calls the Add() function to calculate 1 + 2 and 3 + 4 and output the results
- int Add(int a, int b): This adds up the two integer parameter values, a and b, and returns the calculation result

So, let's create a new C++ project, name it MyCPP_02, and then add a new main.cpp file.

Then, type in the following code for main.cpp:

```cpp
#include <iostream>

int Add(int a, int b)
{
    return a + b;
}

void main()
{
    std::cout << "My Calculations" << std::endl;

    int result = Add(1, 2);
    std::cout << "Integer addition: 1 + 2 = "
                << result
                << std::endl;
    result = Add(3, 4);
    std::cout << "Integer addition: 3+4="
                << result
                << std::endl;

    std::cout << "Finished!";
}
```

The following bullet points will help break down this code:

- The `std:cout` object represents the standard output device, which usually is the output console.

- The `<<` operator is applied to an output stream. In this example, it transfers the value on its right side to the output device on the left side. Multiple `<<` operators can be linked up so that the values can be output in the same order.

- `std:endl` represents a newline character in the output sequence. Another way to insert a newline character into a string is to use `\n` (in C++, the `\n` escape sequence represents a newline character). The output line has the same result as the following substitute – `std::cout << "Integer addition: 3+4=" << result << "\n";`.

- The `int result = Add(1, 2)` statement defines an integer type variable to store the return result calculated by the `Add()` function. C++ data types and variable declarations will be introduced in the next section.

Now, compile and run the program in the pop-up console. You should get the program's output:

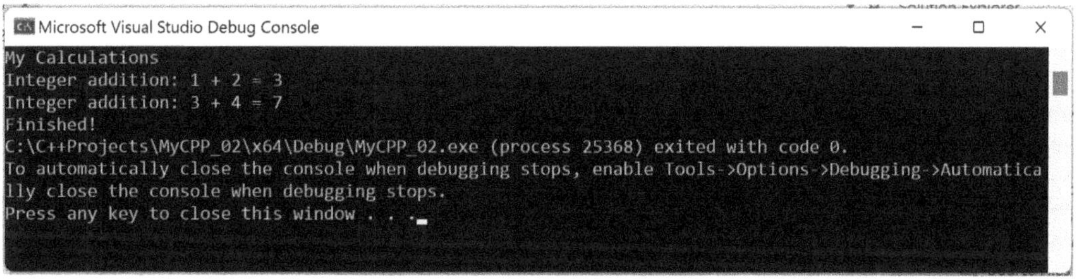

```
Microsoft Visual Studio Debug Console                               —    □    ×
My Calculations
Integer addition: 1 + 2 = 3
Integer addition: 3 + 4 = 7
Finished!
C:\C++Projects\MyCPP_02\x64\Debug\MyCPP_02.exe (process 25368) exited with code 0.
To automatically close the console when debugging stops, enable Tools->Options->Debugging->Automatica
lly close the console when debugging stops.
Press any key to close this window . . .
```

Figure 3.3 – MyCPP_02 calculation output

The program displays `My Calculations` on the first line; after that, it displays the two calculation results on lines 2 and 3, and eventually, it prints out `Finished!`.

In this exercise, we used the `int` data type to declare the `result` variable. Now, let's learn more about C++ data types and variable declarations, user input, operators, and flow control. After that, we will practice using them to improve the `MyCPP_02` calculation project.

Learning the C++ syntax

Learning C++ syntax is crucial to write correct and reliable code, and the ability to write effective C++ code is a fundamental skill that C++ developers must possess. In this section, we will introduce the essentials of C++ syntax, beginning with data types.

Using the C++ data types

Besides the `int` data type that we just have used, C++ has many other built-in data types that can be used. Here is a list of some basic data types:

Data type	Size (bytes)	Description
int	4	Stores signed integer numbers without decimals. Data range: -2,147,483,648 to 2,147,483,647 Example: `int i = -1; int j = 1;`
unsigned int	4	Stores unsigned integer numbers without decimals. Data range: 0 to 4,294,967,295 Example: `unsigned int i = 0; int j = 1;`
float	4	Stores floating-point numbers. Data range: -3.4E+38 to 3.4E+38 Example: `float pi = 3.14f;`
double	8	Stores floating-point numbers with two times the precision of `float`. Data range: -1.7E+308 to 1.7E+308 Example: `double pi = 3.1415926;`
char	1	Stores a single character or a signed integer value that is within the data range of -128 to 127. Example: `char c = 'A';` or `char c = 65;` (The ASCII code assigned to the `'A'` character is 65)
unsigned char	1	Stores a single character or an unsigned integer value that is within the data range of 0 to 255. Example: `char c = 'B';` or `char c = 66;` (The ASCII code assigned to the `'B'` character is 66)
short	2	Stores signed integer numbers without decimals. Data range: -32,768 to 32,767 Example: `short i = -1; short j = 1;`

unsigned short	2	Stores unsigned integer numbers without decimals. Data range: 0 to 65,535 Example: `short i = 0; int j = 65535;`
bool	1	Stores Boolean values with two states: `true` (1) or `false` (0). Example: `bool isAttacking = true;`
void	0	The `void` type indicates the absence of a function's return value.
string	varied	Stores text surrounded by double quotes. The string library needs to be included to use the string type. Example: `#include <string>` `string playerName = "George";`

Figure 3.4 – C++ data types

C++ data types are usually used to declare variables. The next thing we need to learn is how to declare different types of variables.

Defining variables

A **variable** is a container that stores data values. The way to declare (create) a variable is to specify a type, the variable name, and assign a value to it.

Here is the general format to declare a variable:

```
type variableName = value;
```

And here are some examples:

- `int health = 100;`
- `float cash = 50.0f;`
- `bool isHit = false;`
- `string message = "Hello, I am George!";`

You can use commas (`,`) as delimiters to declare multiple variables of the same type. Here is an example that declares three `int` variables, x, y, and z, and stores the 0, 0, and 10 values to them, respectively:

```
int x = 0, y = 0, z = 10;
```

We saw some variable declaration examples, but you may wonder whether there are any rules for variable names. A C++ variable name must be identified with its unique name within a code scope. A variable name can be a short name (such as i, j, x, or y) or a descriptive name (such as playerName, teamId, age, rank, strength, or defense). The general rules for naming variables are as follows:

- Variable names can contain letters, digits, and underscores.

- Variable names must begin with either a letter or an underscore (mostly used for private and protected variables). When coding with C++, it is preferable to follow the **camel casing** standard to name variables (e.g., MyVariables).

- Variable names are case-sensitive.

- Variable names cannot contain whitespace or special characters such as !, @, #, $, and %.

- Variable names cannot use C++ reserved words, such as int, float, and bool.

- For descriptive variable names that contain multiple meaningful words, it is recommended to use caps for the leading letters of each word (it is optional for the first word) and for the rest to be lowercase – for example, playerName or PlayerName.

Putting a const keyword ahead of a variable declaration makes the variable a constant, which means unchangeable and read-only. In C++, once a name is declared to be a constant, this name is equivalent to the stored value – for example, const float PI = 3.14f;. This statement tells the compiler that the PI constant is an equivalent expression of the 3.14f float value and can be used to represent the value in your program.

The benefit of using constant names in your program is that changing the declared value (const float PI = 3.1415926f;, for instance), will update all the places that reference the constant name, PI. It also avoids value inconsistency and reduces the use of system memory.

Using C++ arrays

An **array** is a series of elements of the same type placed in a contiguous memory block that can be individually referenced by the element index. By using an array, you can store a set of similar type values rather than declaring multiple variables for each value.

To declare an array, define the variable type, followed by the name of the array, and then a pair of square brackets. You can provide a number that specifies the length of the array. The following example defines a string array, playerName, that can store eight player names:

```
string playerNames[4];
```

You can initialize array elements by listing the string values. This list of strings should be placed in between a pair of braces and separated with commas. Here is an example of initializing the `playerNames` array:

```
string playerNames[] = { "George", "Sarah", "Willy", "Mike" };
```

Note that the array length is omitted. This is allowed because the compiler will automatically assume the array size to match the number of provided values.

To access the array elements, you can use the array name followed by the bracketed index number. Remember that *the index number starts from 0.* The following example outputs the second player's name and changes the last player's name to `Charles`:

```
Cout << playerNames[1] << endl;
playerNames[4] = "Charles";
```

Using C++ operators

Operators perform operations on variables and values. C++ operators are categorized into five groups:

- **Arithmetic operators**: These perform mathematical operations:

Operator	Operation	Description	Example(s)
+	Addition	Adds together two values	`1 + 2, 2.1 + 3.5`, and `v1 + v2`
-	Subtraction	Subtracts a value from another	`5 - 1, 1.2 - 3.3`, and `v1 - v2`
*	Multiplication	Multiplies two values	`3 * 2, 3.14f * 2`, and `a * b`
/	Division	Divides a value by another	`4 / 2, 3.14f / 2`, and `a / b`
%	Modulus	Returns the remainder of the division	`11 % 3` (returns 2)
++	Increment	Increases the current value by 1	`3++` or `++3` (returns 4)
--	Decrement	Decreases the current value by 1	`3--` or `--3` (returns 2)

Figure 3.5 – Arithmetic operators

- **Assignment operators**: These assign and store values to variables:

Operator	Operation	Description	Example(s)
=	Assign	Assigns a value to the variable	`const float PI = 3.14f;`
+=	Add and assign	Adds a value and assigns the result back	`pi += 3.14f;` (the value of `pi` will be `6.28f`)
-=	Subtract and assign	Subtracts a value and assigns the result back	`pi = PI;` `pi -= 3.14f;` (the value of `pi` will be `0.0f`)
*=	Multiply and assign	Multiplies a value and assigns the result back	`pi = PI;` `pi *= 2.0f;` (the value of `pi` will be `6.28f`)
/=	Divide and assign	Divides a value and assigns the result back	`pi /= 2.0f;` (the value of `pi` will be `1.57f`)
%=	Mod and assign	Mods a value and assigns the result back	`Int n = 13;` `n %= 5;` (the value of `n` will be 3)
&=	Bitwise And and assign	Applies a bitwise And operation first, and then assigns the result back	`char byte = 0b10010001;` `byte &= 0b11100010;` (the value of `byte` will be `0b01000000`; `0b` is a prefix that indicates a binary value expression)
|=	Bitwise Or and assign	Applies a bitwise Or operation first, and then assigns the result back	`char byte = 0b10010001;` `byte |= 0b11100010;` (the value of `byte` will be `0b11100011`)
^=	Bitwise Exclusive Or and assign	Applies a bitwise exclusive Or operation first, and then assigns the result back	`char byte = 0b10010001;` `byte &= 0b11100010;` (the value of `byte` will be `0b01110011`)
<<=	Left-shift and assign	Shifts the value a certain number of bits to the left, and then assigns the result back	`char byte = 0b10010001;` `byte <<= 2;` (the value of `byte` will be `0b01000100`)
>>=	Right-shift and assign	Shifts the value a certain number of bits to the right, and then assigns the result back	`char byte = 0b10010001;` `byte >>= 2;` (the value of `byte` will be `0b00100100`)

Figure 3.6 – Assignment operators

- **Comparison operators**: These compare two values. The results of comparison operators are always a `bool` data type:

Operator	Operation	Description	Example(s) int x = 0, y = 5, z = 0;
==	Equal	Checks whether two values are equal	x == y (false) x == z (true)
!=	Not equal	Checks whether two values are not equal	x != y (true) x != z (false)
>	Greater than	Checks whether the first value is greater than the second value	x > y (false) y > z (true)
<	Less than	Checks whether the first value is less than the second value	x < y (false) z < y (true)
>=	Greater than or equal to	Checks whether the first value is greater than or equal to the second value	x >= y (false) x >= z (true) y >= z (true)
<=	Less than or equal to	Checks whether the first value is less than or equal to the second value	x <= y (true) x <= z (true) y <= z (false)

Figure 3.7 – Comparison operators

- **Logical operators**: These determine the final logical result based on combining two logical statements:

Operator	Operation	Description	Example(s) int x = 0, y = 5, z = 0;
&&	AND	Returns `true` if both statements are true	x > y && y > z (false) x < y && y > z (true)
||	OR	Returns `true` if one statement is true	x > y && y > z (true) x < y && y > z (true)

!	NOT	Returns `true` if the statement is false, and returns `false` if the statement is true	`! (x > y)` (true) `! (x < y)` (false) `! (x == z)` (false) `! (x > y && y > z)` (true) `! (x < y && y > z)` (false)

Figure 3.8 – Logical operators

- **Bitwise operators**: These perform bit-by-bit operations. Binary values are usually used as flags or masks. Bitwise operations are used alongside number types, such as char, short, and int:

Operator	Operation	Description	Example(s) (char b1 = 0b00100101; char b2 = 0b10100011;)
&	Bitwise AND	Compares two values bit by bit. Returns 1 only when both bits are 1; otherwise, it returns 0.	b1 & b2 (0b00100001)
|	Bitwise inclusive OR	Compares two values bit by bit. Returns 0 only when both bits are 0; otherwise, it returns 1.	b1 | b2 (0b10100111)
^	Bitwise exclusive XOR	Compares two values bit by bit. Returns 1 when the two bits are different; otherwise, it returns 0.	b1 ^ b2 (0b10000110)
~	Bitwise NOT	Applies unary complement (bit inversion).	~b1 (0b11011010) ~b2 (0b01011100)
<<	Shift left	Shifts bits left.	b1 << 1 (0b01001010) b1 << 3 (0b00101000) b2 << 2 (0b10001100)
>>	Shift right	Shifts bits right.	b1 >> 1 (0b00010010) b1 >> 3 (0b00000100) b2 >> 2 (0b00101000)

Figure 3.9 – Bitwise operators

Accepting user input

We have already used the `iostream` library's `cout` object to output (print) values. Now, we will learn how to use the `cin` object to accept user input.

`cin` reads data from a keyboard with the extraction operator (`>>`). In the following example, a user can input a number, which is stored in the x variable. Then, the stored value is printed out:

```
int x;
cout << "Please input your age: ";
cin >> x;
cout <<"Your age is: "<< x;
```

This will produce the following output when the user inputs `21`:

```
Please input your age: 21
Your age is: 21
```

Adding C++ comments

Comments are used to provide extra descriptions and explanations to one line or a block of code. Commented text in C++ source files are not treated as executable code, so they are also used to prevent executions of testing code.

Single-line comments start with two forward slashes, `//`, and the entire line of text is treated as a comment, which is not executable code:

```
  string name;    //this is a player name
```

Multiline comments start with `/*` and end with `*/`. All the text lines between these two tags are treated as comment lines, which are not executable code:

```
/* Add(int a, int b) is an addition function
     Parameters: a and b will be added up
     Return value: the addition result
  */
    int Add(int a, int b);
```

We have now learned about lots of elements of functional statements. Since C++ code is executed line by line, we can only use functional statements to write linear programs.

However, what if we want some code blocks to be executed under certain conditions? In such cases, our programs need to become smart and know how to pick the right things to do, rather than follow simple orders. This brings in the need for flow controls.

Controlling the C++ flow

As a complete and powerful programming language, C++ should support not only sequential statements but also conditional branches and repeatable actions. **Flow control** refers to the order of function calls, instructions, and statements that are executed or evaluated while a program runs. It determines which block of code is run under a certain circumstance (condition).

In C++, if, switch, while loop, and for loop statements are mainly used to control program flows.

Working with the if statements

We use the **if statement** to execute a block of C++ code if a condition is true.

An if statement starts from the if keyword followed by a condition, which is enclosed by a pair of parentheses. Following that is the code block that will be run when the condition is true.

The if statement has three forms – if, else, and else if.

if

An if statement checks the condition expression first, and the process is executed only when a condition's result is true:

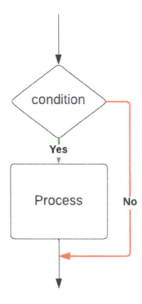

Figure 3.10 – An if statement flowchart

The following code is an example of the `if` statement in use:

```
if(condition)
{
    //Process code block
}
```

else

An `else` statement checks the condition to determine whether the process for `true` (**Process 1**) or the process for `false` (**Process 2**) should be executed:

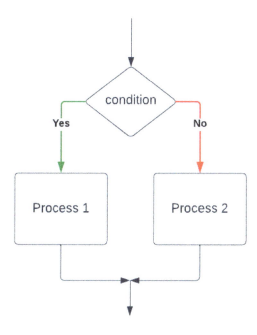

Figure 3.11 – An else statement flowchart

The following code is an example of the `else` statement in use:

```
if(condition)
{
    //Process 1 code block
}
else
{
    //Process 2 code block
}
```

else if

The else if statement allows multiple conditional branches to be sequentially evaluated; each else if branch is checked only if the preceding conditions are false.

As we did for the other two statements, let's have a look at the flowchart for the else if statement:

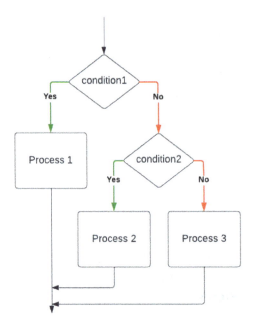

Figure 3.12 – An else if statement flowchart

The following code is an example of the else if statement in use:

```
if(condition1)
{
   //Process 1 code block
}
else if(condition2)
{
   //Process 2 code block
}
else
{
   //Process 3 code block
}
```

Working with the switch statement

The **switch statement** is used to select one of many code blocks to be executed. It evaluates the expression first, then compares the result with the values of each case, and finally, executes the code block (or **process**) when a match is found; otherwise, the default code block is executed:

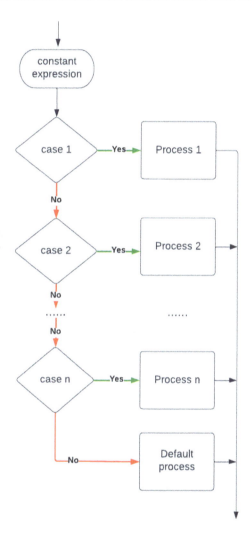

Figure 3.13 – A switch statement flowchart

The following code shows an example of a `switch` statement:

```
switch(expression)
{
case constant1:
  //Process 1 code block
  break;
case constant2:
  //Process 2 code block
  break;
case constantn:
  //Process n code block
  break;
default:
  //Default process code block
}
```

Working with loop statements

Loop statements are used to repeat the execution of a block of code if a specified condition is reached. For example, the calculator program we worked on earlier can be modified to allow users to input multiple data values, calculate them, and output the results.

There are three types of loops:

- The `for` loop
- The `while` loop
- The `do/while` loop

Let's take a look at them.

for loop

When a loop's execution times are known, you can use the `for` loop:

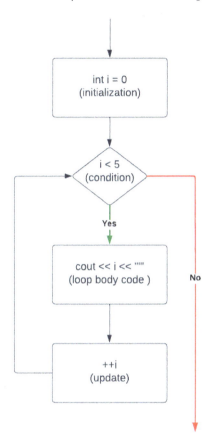

Figure 3.14 – A for loop flowchart

The syntax of the `for` loop is as follows:

```
for(initialization; condition; update)
{
    //loop body code
}
```

Let's break it down:

- The `initialization` statement: This initializes one or more variables. This statement is executed only once at the beginning of the loop.
- The `condition` statement: When this condition is false, the loop ends; otherwise, it continues. The condition is usually related to the value of the initialized variable(s).
- The `update` statement: This updates the value of the initialized variable(s).

Here is an example, which will output `0, 1, 2, 3, 4`:

```
for(int i = 0; i < 5; ++i)
{
   Cout << i << ",";
}
```

while and do/while loops

The `while` and `do/while` loops are similar loop statements. They both loop through a block of code as long as a specified condition is true. The difference is that the former checks the condition before executing the body code, whereas the latter does the check after, which means the `do/while` loop executes the body code at least once, irrespective of whether the condition is true or false.

The `while` loop and `do/while` loop syntaxes are as follows:

while loop example	do/while loop example

<pre>While(condition)
{
   //loop body code
}</pre>
	<pre>do
	{
	   //loop body code
	} while(condition);</pre>

The `condition` statement controls when to exit the loop. When this condition is false, the loop ends; otherwise, the program continues.

The following two examples will output the same results, which are 0, 1, 2, 3 , 4:

while loop example	do/while loop example

<pre>int i = 0;
 while(i < 5)
{
   cout << i << ","
   ++i;
}</pre>
	<pre>int i = 0;
	do
	{
	   Cout << i << ",";
	   ++i;
	} while(i < 5);</pre>

We'll show another pair of examples with different execution results. Here, the do/while loop will output the message executed! once:

while loop example	do/while loop example
| ```
bool condition = false;
while(conditiion)
{
    cout << "executed!"
}
``` | ```
bool condition = false;
do
{
    cout << "executed!"
} while(condition);
``` |

As with the statements, you can see the while and do/while loops illustrated here:

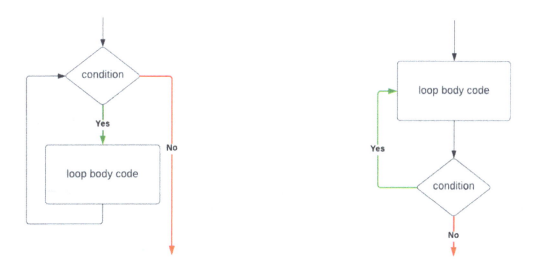

Figure 3.15 – Comparing while loop and do/while loop flowcharts

Using the break and continue statements

The **break statement** is not only used to exit a switch statement but can also be utilized to terminate a loop prematurely.

The **continue statement** halts the current loop and proceeds to the next iteration, bypassing the remaining code within the loop.

Let's look at the following code:

```
while(health > 0)
{
  if(Hit())        //check if hit by others
  {
    health -= 3;  //Reduce health for 3 point
    If(health <= 0)  //check if killed
    {
      Break;  //jump out to DieProcess()
    }
    ProcessHit();  //play hit animation
    Continue;      //start the next iteration
  }
  MoveForward();
}
DieProcess(this);    //play die animation
```

This example shows the following:

- How the break statement terminates the loop when the actor is hit and the health is lower than or equal to 0

- How the continue statement skips moving forward when the actor is hit

With that, we have covered a lot of information about the C++ syntax. In order to practice what we just learned, let's create a new project, MyCPP_03, which is a second version of the calculator program.

Working on the improved calculator program

In this new program, we want to add the following features:

- Allow a user to repeatedly input numbers for calculations

- Make another version of the Add function, which adds float values

- Add some comments to explain what the code blocks do

- Put the Add functions into the separate header and source files – Calculator.h and Calculator.cpp

Follow these steps:

1. Create a new project and name it MyCPP_03.

2. Add the main.cpp, calculator.cpp, and calculator.h files to the project. Your **Solution Explorer** window should now have the files under the Header Files and Source Files folders, like so:

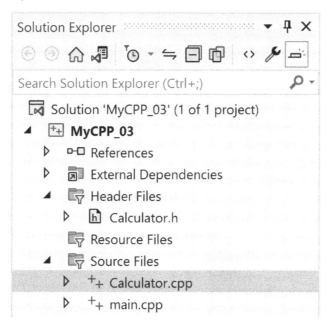

Figure 3.16 – The MyCPP_03 Solution Explorer

3. Next, type the following code into the main.cpp file:

```cpp
#include <iostream>
#include "Calculator.h"

using namespace std;

void main()
{
    cout << "My Calculations" << endl;

    float input1, input2;
    while (true)
    {
        cout << "Input the first value (0 to exit): ";
        cin >> input1;
```

```cpp
            if (input1 == 0)   //exit if the user enters 0
            {
                break;
            }
            cout << "Input the second value (0 to exit): ";
            cin >> input2;
            if (input2 == 0)   //exit if the user enters 0
            {
                break;
            }

            int a = input1;
            int b = input2;
            if (a == input1 && b == input2)
            {
                int result = Add(a, b);
                cout << "Integer addition: "
                    << a << " + " << b << " = "
                    << result
                    << std::endl;
            }
            else
            {
                float result = Add(input1, input2);
                cout << "float addition:  "
                    << input1 << " + " << input2 << " = "
                    << result
                    << std::endl;
            }
        }
    std::cout << "Finished!";
    }
```

Refer to *Figure 3.17*, which depicts the appearance of the main.cpp content in the VS editor.

```
Calculator.h        Calculator.cpp        main.cpp*  ⊡ ✕
MyCPP_03                    ▾    (Global Scope)              ▾   ⬡ main()                       ▾
     1    ⊟#include <iostream>
     2     #include "Calculator.h"
     3
     4     using namespace std;
     5
     6    ⊟void main()
     7     {
     8         cout << "My Calculations" << endl;
     9
    10         float input1, input2;
    11    ⊟    while (true)
    12         {
    13             cout << "Input the first value (0 to exit): ";
    14             cin >> input1;
    15    ⊟        if (input1 == 0)    //exit if the user enters 0
    16             {
    17                 break;
    18             }
    19             cout << "Input the second value (0 to exit): ";
    20             cin >> input2;
    21    ⊟        if (input2 == 0)    //exit if the user enters 0
    22             {
    23                 break;
    24             }
    25
    26             int a = input1;
    27             int b = input2;
    28    ⊟        if (a == input1 && b == input2)
    29             {
    30                 int result = Add(a, b);
    31                 cout << "Integer addition: " << a << " + " << b << " = "
    32                     << result
    33                     << std::endl;
    34             }
    35    ⊟        else
    36             {
    37                 float result = Add(input1, input2);
    38                 cout << "float addition:  " << input1 << " + " << input2 << " = "
    39                     << result
    40                     << std::endl;
    41             }
    42         }
    43         std::cout << "Finished!";
    44     }
```

Figure 3.17 – The main.cpp code

The main.cpp file contains the main() function, which accepts user inputs and outputs the calculation results:

- *Line 1* includes the system's iostream library.

- *Line 2* includes Calculator.h so that the Add functions can be used in this module. The filename is not enclosed with angular brackets – this is usually used for programmer-defined header files (not compiler/IDE-designated directory files, which usually are C++ Standard Library or target platform files).

- *Line 4* uses the namespace `std`. It simplifies using `cout` and `cin` without the namespace tag, `std::`. Namespaces provide a method to prevent name ambiguity in large projects.

- *Line 10* defines two float variables to accept user input.

- *Line 11* is the `while` loop; as the condition is true, this means it is an endless loop.

- *Lines 13 to 41* are the loop body code.

- *Lines 14 and 20* retrieve user inputs and store the values to the two input variables.

- *Lines 15 to 17 and 21 to 24* check whether the user inputs 0 – if it is true, then the loop ends.

- *Lines 26 and 27* trim the decimals from the float values and assign the results to the integer variables.

- *Line 28* checks whether the two input values are integers – if it is true, the `Add(int, int)` function is called; otherwise, the `Add(float, float)` function is called.

4. Then, type the following code into the `Calculator.h` file:

```
#pragma once
/*
   Function Add: adds two integers and returns the result
   Parameters a, b: the two integer input values
*/
int Add(int a, int b);


/*
   Function Add: adds two floats and returns the result
   Parameters a, b: the two float input values
*/
float Add(float a, float b);
```

Refer to the screenshot in *Figure 3.18*, which depicts the appearance of the `Calculator.h` content in the VS editor.

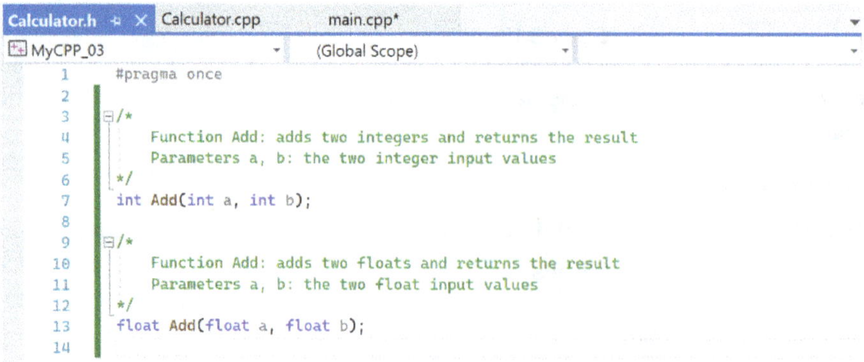

Figure 3.18 – The Calculator.h code

The `Calculator.h` header file only contains the function signature declarations. Whenever other source files need to call these two functions, they have to include this header file with the `#include "Calculator.h"` statement. Let's break down the code:

- *Line 1* includes `#pragma once`, which is a C++ preprocessor directive to ensure the current source file is included only once in a single compilation. We recommend always making it the first line of your header code.

- *Lines 3 to 6 and 9 to 12* are comments that provide more details about the functions.

- *Lines 7 and 13* are the function declarations without the implementations.

5. Then, type the following code into the `Calculator.cpp` file:

```cpp
#include "Calculator.h"

int Add(int a, int b)
{
    return a + b;
}

float Add(float a, float b)
{
    return a + b;
}
```

Refer to the screenshot in *Figure 3.19*, which depicts the appearance of the `Calculator.cpp` content in the VS editor.

Figure 3.19 – The Calculator.cpp code

The `Calculator.cpp` file contains the implementations of the two Add functions. These two functions have the same name but different parameter types – we call this **function overloading**. When the Add function is called, C++ picks the most suitable version. For example, when the two parameters are both float types or one of them is a float type, `Add(float a, float b)` is called. When both of the parameters are integers, `Add(int a, int b)` is called.

6. Now, build and run the program, and then try inputting some data. You should get output as shown in the following figure:

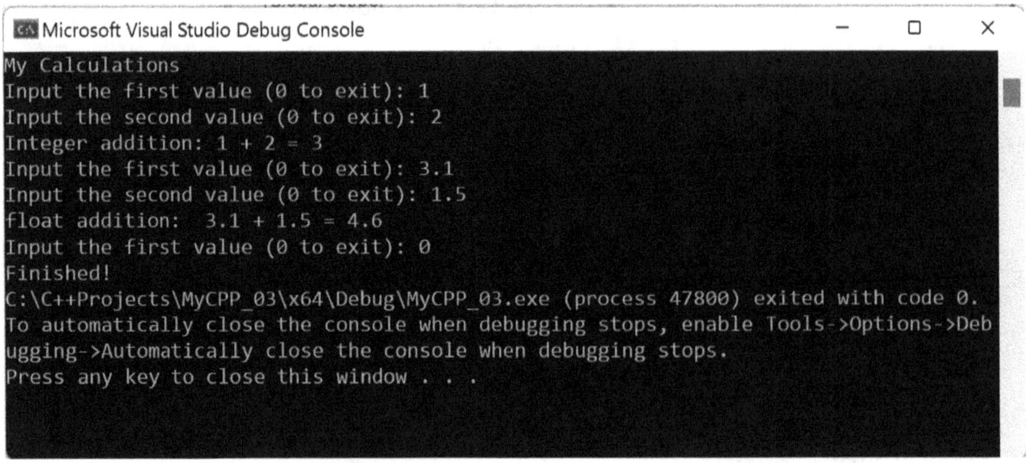

Figure 3.20 – The MyCPP_03 output

We have worked on two versions of the calculator program, which involve variable and function definitions, along with the manipulation of flow control statements. Moving forward, we will learn how to use references and pointers.

Creating references and pointers

When writing C++ code, a variable may not be accessed in only one place. Copying variable values into multiple places brings the risk of inconsistent variable values, as well as lower performance and more memory usage.

Look at the following addition example:

```cpp
float Add(float a, float b)
{
    return a + b;
}


Void main()
{
    int x = 1, y = 2;
    cout << Add(x, y);
}
```

You can see that the function's parameters actually copy the x and y values into the two a and b variables, which means that a and b have their own storage, and any value changes on a and b within the Add function won't affect the values of x and y.

Using references and pointers in C++ can not only help to use less memory and improve performance but also provide the flexibility to modify the original variable values.

References

A **reference** variable is a *reference* to an existing variable, and it is defined with the & operator ahead of the reference variable name. A reference variable name can be considered as a synonym for the original variable name. Once defined, you can use either the reference name or the variable name to indicate the same thing.

Look at the modified addition example:

```cpp
float Add(float &a, float &b)
{
    return a + b;
}

Void main()
{
    int x = 1, y = 2;
    cout << Add(x, y);
}
```

Here, the function parameters are both reference variables. When calling the Add function, the parameter values are not copied; a and b are considered to be the two aliases of x and y at this time. Changing the values of the reference variables inside the Add function will result in changes to the values of x and/or y.

To illustrate the point, here are some more examples:

```cpp
string myName = "George";     //defines a string variable
string &nameOfMe = myName;     //defines a reference
cout << "My name:" << myName;  //outputs "My name: George"
cout << "My name: " << nameOfme;  //outputs "My name: George"
myName = "Li";
cout << "My name:" << nameOfMe;  //outputs "My name: Li"
```

Pointers

A **pointer** is a variable that stores the memory address of another variable, and it is created with the * operator ahead of the variable name. Look at the following modified addition example:

```
float Add(float *a, float *b)
{
   return *a + *b;     //add and return the pointed values
}

Void main()
{
   int x = 1, y = 2;
   cout << Add(&x, &y);   //pass the address of x and y
}
```

Here, the function parameters are both pointers. When calling the Add function, the a and b parameters only copy the input values' memory addresses; to get the pointed values inside the Add function, you have to add the * prefix to the pointer variable names.

Here are some more examples:

```
string myName = "George";    //defines a string variable
string *pMyName = &myName;     //defines a pointer
string *pNameOfMe = pMyName;//defines another pointer
cout << "My name:" << *pMyName;    //outputs "My name: George"
cout << "My name: " << *pNameOfme;//outputs "My name: George"
myName = "Li";
cout << "My name:" << *pMyName;    //outputs "My name: Li"
You just learned cout << "My name:" << *pNameOfMe; //outputs "My name:
Li"
```

Having grasped the significant C++ programming feature of using references and pointers, the next crucial aspect to explore is OOP.

Understanding OOP

Before diving into the world of OOP, it is necessary to lay the groundwork by explaining some fundamental OOP concepts and terms. Following this, you will learn how to create C++ classes and utilize the new classes to instantiate objects, allowing for the practical implementation of OOP principles.

What is OOP?

In the `MyCPP_0x` projects, we wrote functions to perform operations on the data. The approach we used is actually called procedural programming.

OOP is defined as a programming paradigm built on the concept of objects. OPP tries to reflect real-world concepts by creating objects that contain attributes (fields) and functions (methods).

There are three major pillars on which OOP relies:

- **Encapsulation**: This means that data and functions can be wrapped up into classes so that some sensitive data is hidden from users.

- **Inheritance**: This means that a class can derive from another base class to be its child class. The child class can inherit public and protected attributes and functions from the base class. In addition, the child class can also have its own extra attributes and functions.

- **Polymorphism**: This means that one class method can have multiple forms.

The main benefits of using OOP over procedural programming are as follows:

- **Modularity**: Objects act as containers that wrap their attributes and methods up. It eases troubleshooting and collaborative development.

- **Reusability**: Code can be reused through inheritance, which minimizes the possibility of redundant code and change risks.

- **Productivity**: OOP is more productive than procedural programming because of OOP's capabilities for natural world reflection and code reusability.

> **Note**
>
> In addition to the aforementioned points, OOP also has other advantages over procedural programming. You can search online if you are interested in finding out more by yourself (`https://en.wikipedia.org/wiki/Object-oriented_programming`).

What are classes and objects?

Classes and objects are the two main features of OOP. A **class** is a template that can be used to create objects; on the other hand, an **object** is an instance of a class. In other words, a class can be used as a blueprint to instantiate objects.

To illustrate the distinction between classes and objects, let's consider an example. A `Computer` class serves as a blueprint for computer products, encompassing attributes such as `CPUType` and `RAMSize`. This blueprint acts as a template from which a certain number of computers can be manufactured.

In essence, the class defines the common properties and behaviors, whereas the objects represent individual instances that were created based on the class blueprint.

Creating classes in C++

A class is a user-defined data type that starts with the `class` keyword, followed by the class name. The body of the class is defined within a pair of curly braces and terminated by a semicolon at the end.

A class can have attributes, which are variables that represent the properties of the class. It can also have methods (member functions).

Let's create our first class, called `Computer`, which encompasses a private attribute, `_ComesWithMonitor`, and two public attributes, `CPUType` and `RAMSize`:

```
class Computer
{
  Private:
  bool _ComesWithMonitor = true;   //true or false
  Public:
  string CPUType = "Intel";      //"Intel" or "AMD"
  int RAMSize = 4096;        //Unit: Giga bytes

  void TurnOn()
  {
    //...
  }
  void Shutdown()
  {
    //...
  }
  void SetComesWithMonitor(bool ComesWithMonitor)
  {
    _ComesWithMonitor = ComesWithMonitor;
  }
  bool GetComesWithMonitor()
  {
    Return _ComesWithMonitor;
  }
}
```

Let's examine the accessibilities of the attributes and methods defined for the `Computer` class in more detail.

The class attributes

The Computer class has three attributes – _ComesWithMonitor, CPUType, and RAMSize. Note that _ComesWithMonitor is placed under the private group, and the rest of the attributes are placed under the public group.

The public and private specifiers define access scopes for attributes and functions. public attributes and functions can be accessed outside of the class, whereas private attributes and functions are only visible inside the class.

In this example, we used an underscore (_) prefix to indicate that _ComesWithMonitor is a private attribute. However, this coding convention is not obligatory, and you are free to adopt an alternative coding convention of your own.

The class methods

Methods are functions that belong to a class. The Computer class has four functions – TurnOn, Shutdown, SetComesWithMornitor, and GetComesWithMonitor. I only implemented the setter and getter functions, which allow outsiders access to the private flag attribute, _ComesWithMonitor.

Creating objects in C++

Returning to objects, C++ provides two methods to create them.

Method 1 – defining a variable

Here is an example of defining a MyComputer instance from the Computer class:

```
Computer MyComputer;              //method 1
```

This method simply defines a new variable named myComputer. The attributes and the methods of the MyComputer object can be accessed with the dot (.) syntax.

The following example sets the computer's CPUType as "AMD" and the private variable, _ComesWithMonitor, as false by calling the Set function:

```
MyComputer.CPUType = "AMD";
MyComputer.SetComesWithMonitor(false);
```

Method 2 – using the new keyword to instantiate an object and store it to a pointer

Here is an example of using the Computer class to instantiate a new MyComputer instance and store the pointer of the new instance to the pMyComputer pointer:

```
Computer *pMyComputer = new Computer();  //method 2
```

This second method requests the system to allocate a block of memory for the new Computer instance and return the pointer to the memory address. The attributes and the methods of the pMyComputer object can be accessed with the pointer (->) syntax.

The following example also sets the computer's CPUType as "AMD" and the private variable, _ComesWithMonitor, as false by calling the Set function:

```
pMyComputer->CPUType = "AMD";
pMyComputer->SetComesWithMonitor(false);
```

Since the storage of the computer object information is requested by the programmer's code and is dynamically allocated, it is important to release the memory when it is not used. Forgetting to release this memory will cause a memory leak. The delete keyword is used to free unused memory, like so:

```
delete pMyComputer;
```

Based on the knowledge gained in this section, you should now have a solid understanding of implementing OOP classes and objects. Are you enthusiastic about applying this knowledge to convert the MyCPP_03 program into an OOP program? Let's dive into it.

Working on an OOP calculator program

Let's create a new project, MyCPP_04, and create the main.cpp, Calculator.cpp, and Calculator.h files. What we mainly want to do is to create a Calculator class and then make the Add functions the class methods.

Follow these steps:

1. Type the following code into the main.cpp file:

    ```
    #include <iostream>
    #include "Calculator.h"

    using namespace std;

    void main()
    {
        Calculator calculator;   //defines the calculator object
        cout << "My Calculations: " << calculator.GetName() << endl;

        float input1, input2;
        while (true)
    ```

```
    {
        cout << "Input the first value (0 to exit): ";
        cin >> input1;
        if (input1 == 0)  //exit if the user enters 0
        {
            break;
        }

        cout << "Input the second value (0 to exit): ";
        cin >> input2;
        if (input2 == 0)  //exit if the user enters 0
        {
            break;
        }

        int a = input1;
        int b = input2;
        if (a == input1 && b == input2)
        {
            int result = calculator.Add(a, b);
            cout << "Integer addition: " << a << " + " << b << "
= "
                << result
                << std::endl;
        }
        else
        {
            float result = calculator.Add(input1, input2);
            cout << "float addition:  " << input1 << " + " <<
              input2 << " = "
                << result
                << std::endl;
        }
    }

    std::cout << "Finished!";
}
```

Refer to the screenshot in *Figure 3.21*, which depicts the appearance of the main.cpp content in the VS editor.

```cpp
#include <iostream>
#include "Calculator.h"

using namespace std;

void main()
{
    cout << "My Calculations" << endl;

    Calculator calculator;  //defines the calculator object
    float input1, input2;
    while (true)
    {
        cout << "Input the first value (0 to exit): ";
        cin >> input1;
        if (input1 == 0)    //exit if the user enters 0
        {
            break;
        }

        cout << "Input the second value (0 to exit): ";
        cin >> input2;
        if (input2 == 0)    //exit if the user enters 0
        {
            break;
        }

        int a = input1;
        int b = input2;
        if (a == input1 && b == input2)
        {
            int result = calculator.Add(a, b);
            cout << "Integer addition: " << a << " + " << b << " = "
                << result
                << std::endl;
        }
        else
        {
            float result = calculator.Add(input1, input2);
            cout << "float addition:  " << input1 << " + " << input2 << " = "
                << result
                << std::endl;
        }
    }

    std::cout << "Finished!";
}
```

Figure 3.21 – The OOP main.cpp code

The main points to note are as follows:

- *Line 10* creates the `calculator` object
- *Lines 32 and 39* call the calculator's Add method

2. Type the following code into the `Calculator.h` file:

```
#pragma once
#include <iostream>
using namespace std;

class Calculator
{
protected:

  string _name;

public:
  Calculator();     //This is the constructor

  string GetName();

  /*
    Function Add: adds two integers and returns the result
    Parameters a, b: the two integer input values
  */
  int Add(int a, int b);

  /*
    Function Add: adds two floats and returns the result
    Parameters a, b: the two float input values
  */
  float Add(float a, float b);
};
```

Refer to the screenshot in *Figure 3.22*, which depicts the appearance of the `Calculator.h` content in the VS editor.

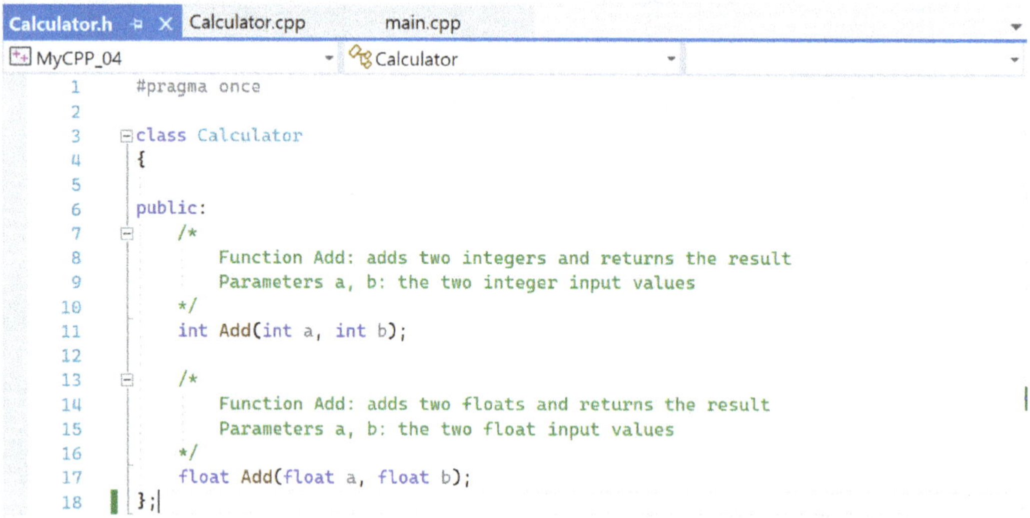

```
Calculator.h  ⇄ ✕   Calculator.cpp        main.cpp
MyCPP_04                          ▼  Calculator                ▼
    1        #pragma once
    2
    3     ⊟ class Calculator
    4       {
    5
    6       public:
    7     ⊟     /*
    8                 Function Add: adds two integers and returns the result
    9                 Parameters a, b: the two integer input values
   10             */
   11             int Add(int a, int b);
   12
   13     ⊟     /*
   14                 Function Add: adds two floats and returns the result
   15                 Parameters a, b: the two float input values
   16             */
   17             float Add(float a, float b);
   18       };
```

Figure 3.22 – The OOP Calculator.h code

The main points to note are as follows:

- *Line 3* defines the `Calculator` class

- *Lines 5 to 17* are the body of the class, which has the two methods

3. Type the following code into the `Calculator.cpp` file:

```cpp
#include "Calculator.h"

Calculator::Calculator()
{
    _name = "Addition Calculator";
}

string Calculator::GetName()
{
    return _name;
}

int Calculator::Add(int a, int b)
{
    return a + b;
}
```

```
float Calculator::Add(float a, float b)
{
    return a + b;
}
```

Refer to the screenshot in *Figure 3.23*, which depicts the appearance of the Calculator.cpp content in the VS editor.

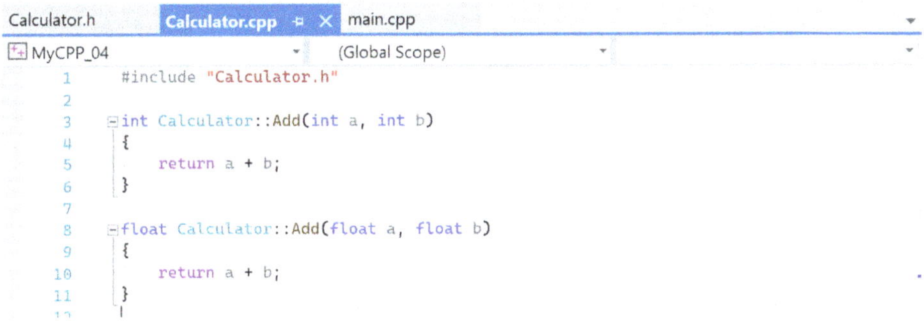

| Calculator.h | **Calculator.cpp** ⚓ ✕ | main.cpp |

```
MyCPP_04                        (Global Scope)
  1      #include "Calculator.h"
  2
  3    ⊟int Calculator::Add(int a, int b)
  4     │{
  5     │    return a + b;
  6     │}
  7
  8    ⊟float Calculator::Add(float a, float b)
  9     │{
 10     │    return a + b;
 11     │}
```

Figure 3.23 – The OOP Calculator.cpp code

Let's break this down:

- *Lines 3 to 6* show the implementation of the integer version's Add method
- *Lines 8 to 11* show the implementation of the float version's Add method

In this example, the two Add methods of the class are implemented outside of the definition of the class. To make a function a method of a class, we use the name of the class followed by the :: operator ahead of the function names.

Build and run this modified program. You should get the same result as that obtained from the *Working on the improved calculator program* exercise (see *Figure 3.20*).

At this point, the calculator is functioning as anticipated. Building upon this progress, we will explore additional aspects of OOP, such as class constructors, getter functions, and class extension, to further expand our knowledge of OOP.

Adding constructor and getter functions for the calculator class

A **constructor** of a class is a special method that is automatically called when an object is created. A constructor's name must be the same as the class name, and the function should have no return type. Developers usually put the class attribute initializations in the constructors.

Let's add a constructor for the Calculator class:

1. Add a private _name attribute to the Calculator class, which will store the calculator's name:

    ```
    string _name;
    ```

2. Add the constructor and the getter declarations to the class:

```cpp
Calculator();
String GetName();
```

3. Implement the constructor and the getter in `Calculator.cpp`:

```cpp
Calculator::Calculator()
{
  _name = "Addition Calculator";   //set the name
}

string Calculator::GetName()
{
  Return _name;
}
```

Having successfully added the constructor and the GetName function to the Calculator class, let's now explore extending the Calculator class by creating a new subclass named CalculatorEx.

Creating the CalculatorEx class, which inherits from the Calculator class

The final aspect of OOP that we want to experiment with is inheritance.

By now, the Calculator class only offers additions, which may be good enough under certain circumstances, but what if we need the calculator to support subtraction?

Directly adding the subtraction functions to the Calculator class may solve the problem, but when certain applications do not need subtractions, the new subtraction functions become unused code.

The better way to solve the problem is to create a second class, which can be named CalculatorEx. This class inherits from the Calculator class and has its own subtraction methods. This makes it possible to use the right version of the calculator for different situations.

Let's use a UML class diagram to illustrate the relationship between the two classes (see *Figure 3.24*).

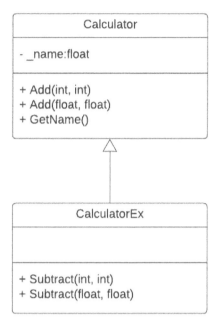

Figure 3.24 – A CalculatorEx and Calculator class diagram

UML

UML stands for **Unified Modeling Language**. A **UML class diagram** is a graphical representation to construct and visualize object-oriented systems.

In the diagram, note the following notations:

- The upward arrow indicates that `CalculatorEx` inherits from `Calculator`
- The - notation indicates private or protected attributes or functions
- The + notation indicates public attributes or functions

To implement the class inheritance, open the MyCPP_04 project, add a new file called CalculatorEx.h, and then enter the following code:

```
#pragma once

#include "Calculator.h"

class CalculatorEx : public Calculator
{
public:

    CalculatorEx();        //This is the constructor

    int Subtract(int a, int b);

    float Subtract(float a, float b);
};
```

Refer to the screenshot in *Figure 3.25*, which depicts the appearance of the CalculatorEx.h content in the VS editor.

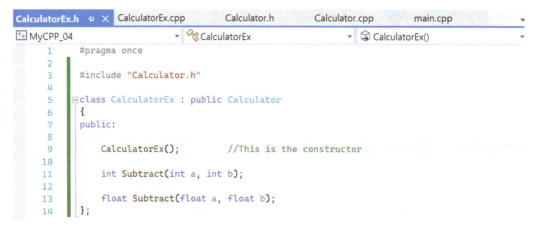

Figure 3.25 – The OOP CalculatorEx.h code

Let's break down this code:

- *Line 9* is the constructor's declaration

- *Lines 11 and 13* are the declarations of the two overloaded Subtract functions

Then, create another new file, called `CalculatorEx.cpp`, and enter this code:

```cpp
#include "CalculatorEx.h"

CalculatorEx::CalculatorEx()
{
  _name = "Advanced Calculator";
}

int CalculatorEx::Subtract(int a, int b)
{
  return a - b;
}

float CalculatorEx::Subtract(float a, float b)
{
  return a - b;
}
```

Refer to the screenshot in *Figure 3.26*, which depicts the appearance of the `CalculatorEx.cpp` content in the VS editor.

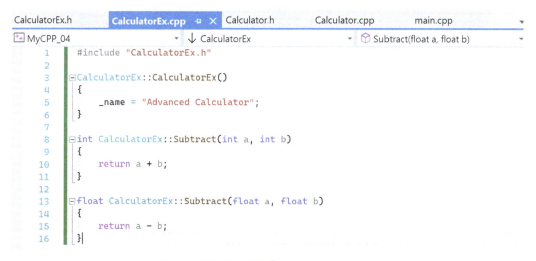

Figure 3.26 – The OOP CalculatorEx.cpp code

Let's break down this code:

- *Lines 3 to 6* are the constructor of `CalculatorEx`, which will be executed after the constructor of `Calculator`, which is the base class

- *Lines 8 to 11* and *lines 13 to 16* are the implementations of the two overloaded `Subtract` functions

One more thing that needs to be done is to move the `_name` variable in the `Calculator` class from the `private` group to the `protected` group. The difference is that private attributes and methods are hidden from their child classes, whereas protected attributes and methods are like public ones that are inherited by the child classes:

```
Protected:
    String _name;
```

Since the `CalculatorEx` class is a child class of the `Calculator` class, it inherits all the attributes and functions from its parent class (the `_name` attribute and the `Add` functions) and has its own unique functions (the `Subtract` functions). It is obvious that `CalculatorEx` is more powerful than `Calculator`.

Would you like to integrate the `CalculatorEx` class into the `Calculator` program? We encourage you to try for yourself and explore further on your own. However, we will still provide the `MyCPP_05` project as a reference, which you can download from the GitHub repository.

Summary

In this chapter, you learned the essentials of the C++ programing language. This chapter covered the topics of the compilation process, the program structure, data types, variable creation, functions, comments, the Standard Library's user input, reference and pointer creations, flow control, and OOP.

The three exercises should be very helpful to learn and practice the C++ syntax, procedural programming, and OOP skills. The five `MyCPP_x` projects are also good samples that can be referenced in your later studies.

Since C++ is a very powerful programming language, providing ample features and functionalities, it is not possible to cover everything in this chapter. We will explain other C++ syntaxes when they are used.

You should now have the necessary knowledge about C++ programming. In the next chapter, we will investigate the Unreal Engine-generated C++ source code for the shooter game. This should help you to quickly understand the commonly used Unreal Engine classes and APIs.

4

Investigating the Shooter Game's Generated Project and C++ Code

Game developers choose to use game engines because they provide toolsets that help accelerate developer workflows. In Unreal Engine, what C++ scripting does is to program based on Unreal Engine's predefined classes and APIs. Therefore, the best way to start learning about and understanding Unreal Engine C++ scripting is to review an engine-generated project.

By investigating the project structure and the source code, you will obtain not only an overall view of the C++ project's structure but also the most used engine programming APIs. Based on the knowledge you just gained and the MyShooter C++ project created in *Chapter 1*, we will go through the following topics in this chapter:

- Understanding the MyShooter C++ project structure
- Understanding the game program structure
- Getting familiar with the source code
- Launching Unreal Editor and opening the game project in Visual Studio

Technical requirements

In this chapter, while examining the code, we will just be previewing the concepts and the use of APIs, elements, and functions in the project. The technical details will be introduced in the subsequent chapter.

Understanding the MyShooter C++ project structure

You already learned about the regular C++ project structure in *Chapter 3*; now, let's take a close look at an Unreal game's C++ project structure.

Open **MyShooter** in Unreal Editor and select the C++ Classes/MyShooter folder in the **Content Drawer** window. There, you can find five C++ class files:

Figure 4.1 – MyShooter C++ project source files

Double-click on any C++ class file to open the C++ project in VS.

Now, let's look at **Solution Explorer**:

Figure 4.2 – MyShooter C++ project Solution Explorer

The first layer of the tree has three folder nodes – `Engine`, `Games`, and `Visualizers`:

- The `Engine` folder contains the **Unreal Engine** project and the source code. In some cases, you may want to modify and customize the engine code.

- The `Games` folder contains game projects. Here, it only has one project – `MyShooter`.

- The `Visualizers` folder contains the `.natvis` files, which contain XML syntax configurations for viewing objects and other data in Visual Studio. You can customize data visualizations in the `.natvis` files; this means that users can view data better (as per their preferences) when editing or debugging code.

For now, we are only interested in the structure of `MyShooter`. This folder has all the files belonging to the game project (see *Figure 4.3*). Developers usually add new source files to the project under the `\Source` subfolder.

Figure 4.3 – MyShooter C++ project structure

Let's look at the project structure in detail:

- The `References` folder contains references to other solutions or shared projects. We are not going to do anything on references in this book, so keep it empty.

- The `External Dependencies` folder has all the dependencies' header files that you will need when writing the game code. Files under this folder are automatically generated by **IntelliSense**. You can edit the project's `MyShooter.build.cs` build file to tell IntelliSense what dependencies you need in this project. By default, the `Core`, `CoreUObject`, `Engine`, and `InputCore` modules are added to the public dependency list.

- The `Source` folder has the `MyShooter.uproject` Unreal project files and two subfolders under it:

  - The `\Config` subfolder has all the `.ini` configuration files.

  - The `\MyShooter` subfolder, which has an identical name to the Unreal project, contains the build and target files (these are **C#** `.cs` files). All the generated C++ source code files (`.cpp` and `.h` files) are placed under the `\MyShooter` subfolder.

Now that we have a basic idea of the Unreal C++ project structure, let's explore the source code inside the `\MyShooter` folder. We will begin by introducing the basic game program structure, and then look into the C++ source code.

Understanding the game program structure

A typical and simple game program usually has three phases – **game initialization**, **game loop**, and **game end** – and Unreal Engine handles these internally.

Unreal Engine offers a range of programming interfaces, including base classes, APIs, and systems, empowering developers to create highly immersive and interactive games. Leveraging these interfaces, developers write C++ code that integrates their custom functionality into the game.

The following flowchart illustrates the fundamental game flow and should give you an idea about writing game code for functional modules:

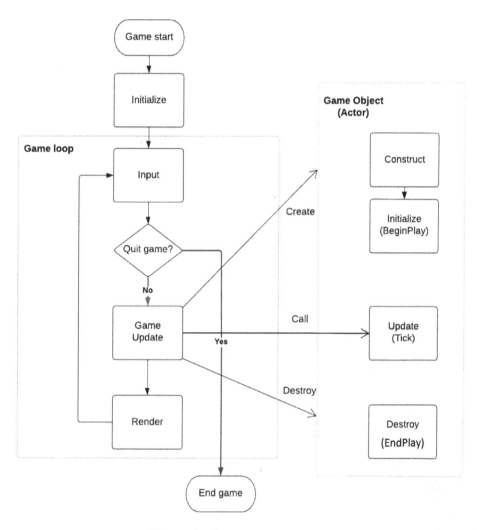

Figure 4.4 – Game program structure

Let's take a look at some of these ideas:

- **Game initialization** happens when a game program starts. All the initialization tasks, such as setting the display mode, loading required contents, spawning game objects, and so on, are done during this phase.

- **Game loop** is the main gameplay period. A game loop involves the repeated retrieval of player inputs, updating of the game, and drawing of the game scene. A complete game loop also refers to the accomplishment of one frame update.

While updating the game, game objects (actors) – bullets that are being shot, for instance – may be created. The created actor also needs to be initialized. Unreal actors have their own `OnConstruction()` and `BeginPlay()` event functions. These two functions are only called once, when the object is created. The actor's `Tick()` event function is called in every frame. Game logic and controls are handled inside this function.

- **Game end** is a phase that ends the game. It takes care of cleanup work, such as destroying existing actors, releasing memory, shutting down network connections, and so on. The actor's `EndPlay()` event function is called at this time. You need to do the cleanup work for the actor when this function gets called.

The actor event functions have corresponding Blueprint Event Nodes:

Actor Blueprint Event	Actor C++ Event Function	Comment
 Construction Script	`OnConstruction()`	Called when the actor is created
 Event BeginPlay	`BeginPlay()`	Called after `OnConstruction()` when the actor is created
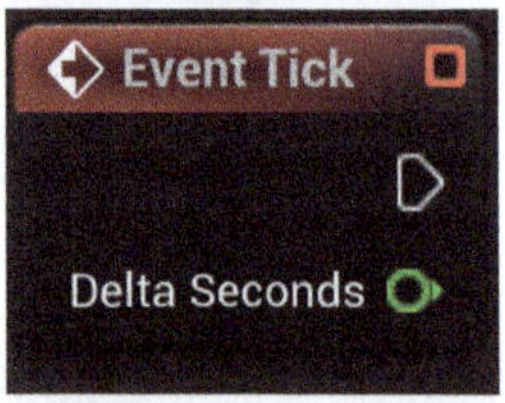 Event Tick / Delta Seconds	`Tick()`	Called in every frame
 Event End Play / End Play Reason	`EndPlay()`	Called before the actor is destroyed

Figure 4.5 – An actor's C++ event functions and the corresponding Blueprint Event Nodes

Not all of the event functions are mandatory for actor classes. In `MyShooterCharacter.h`, only the `BeginPlay()` function is declared and the implementation can be found in `MyShooterCharacter.cpp`.

Now it is time to investigate the generated C++ source code files.

Getting familiar with the source code

In this section, we will go through the generated .cpp and .h source code and explain the code lines or line blocks so that you can better understand how code is organized and written to implement the expected functionalities.

First of all, let's take a look at the `MyShooterCharacter.h` header file for the definition of a typical Unreal class.

MyShooterCharacter.h

`MyShooterCharacter.h` is the header file that defines the game's character class. The `MyShooterCharacter` class is going to be used to create player characters.

The code of this header file can be divided into four parts:

- The definition of the `AMyShooterCharacter` class
- The definition of the class variables
- The declaration of the class member functions
- The declaration of the functions for setting up inputs and the getter functions

The header file's first part mainly defines the `MyShooterCharacter` class, which inherits from Unreal's `ACharacter` class:

```
MyShooterCharacter.h  ⇄  ✕  MyShooterCharacter.cpp
MyShooter                              ▼    (Global Scope)                        ▼                                ▼
    1        // Copyright Epic Games, Inc. All Rights Reserved.
    2
    3        #pragma once
    4
    5      ⊟#include "CoreMinimal.h"
    6       #include "GameFramework/Character.h"
    7       #include "MyShooterCharacter.generated.h"
    8
    9        class UInputComponent;
   10        class USkeletalMeshComponent;
   11        class USceneComponent;
   12        class UCameraComponent;
   13        class UAnimMontage;
   14        class USoundBase;
   15
   16      ⊟// Declaration of the delegate that will be called when the Primary Action is triggered
   17       // It is declared as dynamic so it can be accessed also in Blueprints
   18       DECLARE_DYNAMIC_MULTICAST_DELEGATE(FOnUseItem);
   19
   20       UCLASS(config=Game)
   21      ⊟class AMyShooterCharacter : public ACharacter
   22       {
   23           GENERATED_BODY()
```

Figure 4.6 – MyShooterCharacter.h (Part 1)

The code is broken down as follows:

- *Lines 9-14* declare some Unreal classes that will be used in this module. C++ allows the use of external classes without implementing them inside a module, which means that these classes are implemented somewhere else.

- *Line 18* is a C++ macro offered by Unreal Engine that can be used to declare the FOnUseItem delegate function type. This delegate function type is then used to define the OnUseItem delegate function variable to which external functions can be dynamically assigned. In the MyShooter example, the weapon will attach its Fire() function to this delegate function variable so that when the player pushes the **Fire** button, the delegated function is called.

> **Note**
>
> A **delegate** is essentially a function pointer that can be dynamically bound to one or more methods or functions. Delegates provide a way to invoke multiple functions when a particular event or action occurs. You can refer to the Unreal Engine documentation for more information: https://docs.unrealengine.com/5.0/en-US/delegates-and-lamba-functions-in-unreal-engine/.

- *Line 20* uses a UCLASS macro to indicate that the C++ class defined next will be part of Unreal's **Reflection System** (Unreal's internal system that interprets information, such as macros and specifiers, to process associated C++ classes, variables, and functions). UCLASSes are recognized by Unreal Editor and can use the engine's memory management system. The UCLASS macro on *line 20* also has a specifier, config=game, which indicates that this class is allowed to save its data in the DefaultGame.ini configuration file.

- *Line 21* defines AMyShooterCharacter, which inherits from the engine's ACharacter class. The class has the prefix A, which is required by the engine's coding standard; all classes that inherit from AActor are prefixed with A. Since ACharacter inherits from APawn, APawn inherits from AActor, and AMyShooterCharacter is a child class of ACharacter, they all should be prefixed with A.

- The GENERATED_BODY macro on *line 23* should always be placed on the first line of the class definition.

The second part of the header file defines the class variables (properties) and declares the class constructor and the `BeginPlay()` function:

```
MyShooterCharacter.h  ╺ ✕  MyShooterCharacter.cpp
 MyShooter                              ▾  AMyShooterCharacter
    21    ⊟class AMyShooterCharacter : public ACharacter
    22     {
    23         GENERATED_BODY()
    24
    25         /** Pawn mesh: 1st person view (arms; seen only by self) */
    26         UPROPERTY(VisibleDefaultsOnly, Category=Mesh)
    27         USkeletalMeshComponent* Mesh1P;
    28
    29         /** First person camera */
    30         UPROPERTY(VisibleAnywhere, BlueprintReadOnly, Category = Camera, meta = (AllowPrivateAccess = "true"))
    31         UCameraComponent* FirstPersonCameraComponent;
    32
    33     public:
    34         AMyShooterCharacter();
    35
    36     protected:
    37         virtual void BeginPlay();
    38
    39     public:
    40         /** Base turn rate, in deg/sec. Other scaling may affect final turn rate. */
    41         UPROPERTY(VisibleAnywhere, BlueprintReadOnly, Category=Camera)
    42         float TurnRateGamepad;
    43
    44         /** Delegate to whom anyone can subscribe to receive this event */
    45         UPROPERTY(BlueprintAssignable, Category = "Interaction")
    46         FOnUseItem OnUseItem;
```

Figure 4.7 – MyShooterCharacter.h (Part 2)

The code is broken down as follows:

- *Lines 27, 31, 42, and 46* define four variables: `Mesh1P`, `FirstPersonCameraComponent`, `TurnRateGamepad`, and `OnUseItem`. The `USkeletonMeshComponent` and `UCameraComponent` engine classes both inherit from the `UObject` class, so they are prefixed with `U`.

- The `UPROPERTY` macros on *lines 26, 29, 41, and 45* are used to expose the variables defined on the next lines to Unreal Editor and include the properties in the engine's memory management system. `UPROPERTY` can have specifiers, such as `VisibleAnywhere`, `BlueprintReadOnly`, `Category`, and so on. (We will explain the specifiers after we have finished reviewing the header file code.)

- *Line 34* is the declaration of the class's constructor, which has no return type and has the same name as the class. The constructor is called when the class object is created.

- *Line 37* declares the `BeginPlay()` function that is called during the object creation time after the constructor. The `virtual` keyword indicates that this function can be overridden by its subclasses.

- The `public:` keyword on *lines 33 and 39* indicates that the following variables and functions can be accessed either internally or externally.

- The `protected:` keyword on *line 36* indicates that the following variables and functions can only be accessed internally and are also visible to the subclasses.

- *Lines 25-31* have no access keyword above them, so they will be allied with the `private:` keyword, which means internal access only.

The third part of the header file declares functions that handle character moves and player touch inputs:

```
MyShooterCharacter.h  ┼  ×   MyShooterCharacter.cpp
MyShooter                              ▼  AMyShooterCharacter                              ▼
    46          FOnUseItem OnUseItem;
    47     protected:
    48
    49          /** Fires a projectile. */
    50          void OnPrimaryAction();
    51
    52          /** Handles moving forward/backward */
    53          void MoveForward(float Val);
    54
    55          /** Handles strafing movement, left and right */
    56          void MoveRight(float Val);
    57
    58          /**
    59           * Called via input to turn at a given rate.
    60           * @param Rate  This is a normalized rate, i.e. 1.0 means 100% of desired turn rate
    61           */
    62          void TurnAtRate(float Rate);
    63
    64          /**
    65           * Called via input to turn look up/down at a given rate.
    66           * @param Rate  This is a normalized rate, i.e. 1.0 means 100% of desired turn rate
    67           */
    68          void LookUpAtRate(float Rate);
    69
    70          struct TouchData { ... };
    78          void BeginTouch(const ETouchIndex::Type FingerIndex, const FVector Location);
    79          void EndTouch(const ETouchIndex::Type FingerIndex, const FVector Location);
    80          void TouchUpdate(const ETouchIndex::Type FingerIndex, const FVector Location);
    81          TouchData   TouchItem;
```

Figure 4.8 – MyShooterCharacter.h (Part 3)

The code is broken down as follows:

- *Lines 50, 53, 56, 62, 68, and 78-80* are function declarations.

- *Line 70* defines the `TouchData` structure (called **struct** in C++). A C++ `struct` groups related variables in one place. Each variable in a `struct` is called a member of the `struct`. Structs are defined in a manner similar to classes, except for the use of the `struct` keyword instead of `class`. By default, `struct` variables and functions allow public access, making them accessible externally.

- *Line 81* defines the `TouchData` variable.

The last part of the `MyShooterCharacter.h` header file mainly includes the input setup and the getter functions:

```
MyShooterCharacter.h  ⇆ ✕  MyShooterCharacter.cpp
MyShooter                                              AMyShooterCharacter
  82
  83    protected:
  84        // APawn interface
  85        virtual void SetupPlayerInputComponent(UInputComponent* InputComponent) override;
  86        // End of APawn interface
  87
  88        /*
  89         * Configures input for touchscreen devices if there is a valid touch interface for doing so
  90         *
  91         * @param   InputComponent  The input component pointer to bind controls to
  92         * @returns true if touch controls were enabled.
  93         */
  94        bool EnableTouchscreenMovement(UInputComponent* InputComponent);
  95
  96    public:
  97        /** Returns Mesh1P subobject **/
  98        USkeletalMeshComponent* GetMesh1P() const { return Mesh1P; }
  99        /** Returns FirstPersonCameraComponent subobject **/
 100        UCameraComponent* GetFirstPersonCameraComponent() const { return FirstPersonCameraComponent; }
 101
 102    };
```

Figure 4.9 – MyShooterCharacter.h (Part 4)

The code is broken down as follows:

- *Line 85* overrides its APawn base class's virtual function named `SetupPlayerInputComponent`, which is called by the engine during the setting-up phase.

- The getter functions on *lines 98 and 100* provide public functions so that other objects can retrieve the pointer of the private variables.

When reviewing the `MyShooterCharacter.h` code, the UPROPERTY specifiers were not explained. Property specifiers are added to control how the property behaves with various aspects of the engine and editor. The following are the specifiers used in the `MyShooterCharacter.h` file:

- `VisibleDefaultsOnly` indicates that the property is only visible in property windows for archetypes, and cannot be edited

- `VisibleAnywhere` indicates that this property is visible in all property windows, but cannot be edited

- `BlueprintReadOnly` indicates that this property is readable but not editable to Blueprints

- `BlueprintAssignable` indicates that Blueprints can assign a value to this property

- `Category=` specifies the category of the property when displayed in Blueprint editing tools

More information about property specifiers can be found in Unreal Engine's official online documentation: `https://docs.unrealengine.com/5.0/en-US/unreal-engine-uproperty-specifiers/`

MyShooterCharacter.cpp

Functions declared in the MyShooterCharacter.h file should be implemented in the MyShooterCharacter.cpp file.

The first implemented function is the class constructor:

```
MyShooterCharacter.h        MyShooterCharacter.cpp  ⊕ ×                                    TP_WeaponComponent.cpp  ⊠ ×  ▾
⊡ MyShooter                              ▾  ↓ AMyShooterCharacter              ▾  ⊙ AMyShooterCharacter()                   ▾
    12    ⊟//////////////////////////////////////////////////////////////////////
    13      // AMyShooterCharacter
    14
    15    ⊟AMyShooterCharacter::AMyShooterCharacter()
    16      {
    17          // Set size for collision capsule
    18          GetCapsuleComponent()->InitCapsuleSize(55.f, 96.0f);
    19
    20          // set our turn rates for input
    21          TurnRateGamepad = 45.f;
    22
    23          // Create a CameraComponent
    24          FirstPersonCameraComponent = CreateDefaultSubobject<UCameraComponent>(TEXT("FirstPersonCamera"));
    25          FirstPersonCameraComponent->SetupAttachment(GetCapsuleComponent());
    26          FirstPersonCameraComponent->SetRelativeLocation(FVector(-39.56f, 1.75f, 64.f)); // Position the camera
    27          FirstPersonCameraComponent->bUsePawnControlRotation = true;
    28
    29          // Create a mesh component that will be used when being viewed from a '1st person' view (when controlling this pawn)
    30          Mesh1P = CreateDefaultSubobject<USkeletalMeshComponent>(TEXT("CharacterMesh1P"));
    31          Mesh1P->SetOnlyOwnerSee(true);
    32          Mesh1P->SetupAttachment(FirstPersonCameraComponent);
    33          Mesh1P->bCastDynamicShadow = false;
    34          Mesh1P->CastShadow = false;
    35          Mesh1P->SetRelativeRotation(FRotator(1.9f, -19.19f, 5.2f));
    36          Mesh1P->SetRelativeLocation(FVector(-0.5f, -4.4f, -155.7f));
    37
    38      }
```

Figure 4.10 – MyShooterCharacter.cpp (the class constructor)

Let's look at the code:

- *Line 18* gets ACharacter's capsule component and then initializes the size for collision detections by calling its InitCapsuleSize method.

- *Line 21* sets the initial value for the TurnRateGamepad variable.

- *Line 24* uses the CreateDefaultSubobject function to create the player camera and stores the camera component pointer to the FirstPersonCameraComponent variable. CreateDefaultSubObject is a template function that has the UCameraComponent type enclosed by a pair of angle brackets. This makes the function create an instance of UCameraComponent and return the pointer to the new instance.

- *Lines 25-27* set the `FirstPersonCameraComponent` properties.

- *Lines 30-36* do similar work to creating and setting up the camera, with the `SkeletalMeshComponent` created and initialized here.

The next function is the `MyShooterCharacter` `BeginPlay()` function, which simply calls its `ACharacter` base class's `BeginPlay()` function (see *Figure 4.11*):

```
40    void AMyShooterCharacter::BeginPlay()
41    {
42        // Call the base class
43        Super::BeginPlay();
44
45    }
```

Figure 4.11 – MyShooterCharacter.cpp (the BeginPlay function)

Even though `MyShooterCharacter` has nothing to do here, its `ACharacter` base class may do some meaningful work, so don't forget to call the overridden function.

The next function is the `SetupPlayerInputComponent` function, which binds player input actions and axis changes to their handling functions (see *Figure 4.12*). For example, when the **Jump** button is pressed, the `Jump` function within the `ACharacter` class is called.

```
49    void AMyShooterCharacter::SetupPlayerInputComponent(class UInputComponent* PlayerInputComponent)
50    {
51        // Set up gameplay key bindings
52        check(PlayerInputComponent);
53
54        // Bind jump events
55        PlayerInputComponent->BindAction("Jump", IE_Pressed, this, &ACharacter::Jump);
56        PlayerInputComponent->BindAction("Jump", IE_Released, this, &ACharacter::StopJumping);
57
58        // Bind fire event
59        PlayerInputComponent->BindAction("PrimaryAction", IE_Pressed, this, &AMyShooterCharacter::OnPrimaryAction);
60
61        // Enable touchscreen input
62        EnableTouchscreenMovement(PlayerInputComponent);
63
64        // Bind movement events
65        PlayerInputComponent->BindAxis("Move Forward / Backward", this, &AMyShooterCharacter::MoveForward);
66        PlayerInputComponent->BindAxis("Move Right / Left", this, &AMyShooterCharacter::MoveRight);
67
68        // We have 2 versions of the rotation bindings to handle different kinds of devices differently
69        // "Mouse" versions handle devices that provide an absolute delta, such as a mouse.
70        // "Gamepad" versions are for devices that we choose to treat as a rate of change, such as an analog joystick
71        PlayerInputComponent->BindAxis("Turn Right / Left Mouse", this, &APawn::AddControllerYawInput);
72        PlayerInputComponent->BindAxis("Look Up / Down Mouse", this, &APawn::AddControllerPitchInput);
73        PlayerInputComponent->BindAxis("Turn Right / Left Gamepad", this, &AMyShooterCharacter::TurnAtRate);
74        PlayerInputComponent->BindAxis("Look Up / Down Gamepad", this, &AMyShooterCharacter::LookUpAtRate);
75    }
```

Figure 4.12 – SetPlayerInputComponent in MyShooterCharacter.cpp

All the input **Action Mappings** (`Jump` and `PrimaryAction`) and the **Axis Mappings** (`move forward / backward`, for example) are defined in the engine editor settings. In the engine editor, open the **Project Settings** window and select **Engine | Input**, and you should see the input map:

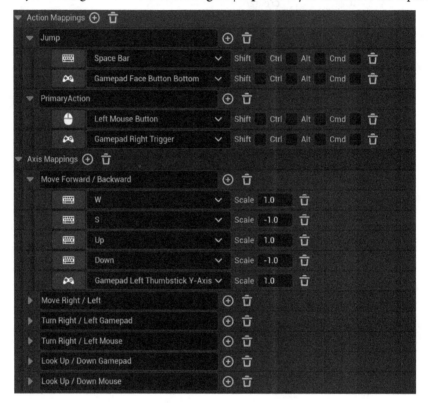

Figure 4.13 – MyShooter project editor input map

If you pay attention to *line 59* in *Figure 4.12*, you will notice that the `OnPrimaryAction` function is bound to `PrimaryAction`. When the left mouse button is clicked or the gamepad's right trigger is pressed, the `OnPrimaryAction` function is called. Then, the `OnPrimaryAction` function uses the delegated variable to broadcast calling functions assigned to it:

```
77    void AMyShooterCharacter::OnPrimaryAction()
78    {
79        // Trigger the OnItemUsed Event
80        OnUseItem.Broadcast();
81    }
```

Figure 4.14 – The OnPrimaryAction function in MyShooterCharacter.cpp

The remaining functions in `MyShooterCharacter.cpp` are responsible for character controls. We are not going to go deep into the implementation details. It is recommended that you read the code and use it as a reference in your future development work.

Since `AMyShooterCharacter` is a UCLASS, it can be inherited by not only C++ subclasses but also Blueprints. The MyShooter game has a `BP_FirstPersonCharacter` Blueprint, which inherits from the `AMyShooterCharacter` class:

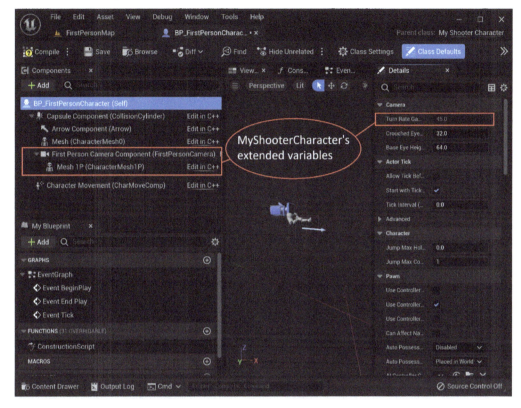

Figure 4.15 – BP_FirstPersonCharacter

From here, you can see that most properties are derived from the `ACharacter` class and only three new properties are added to the new class.

Note that the `FirstPersonCamera` and `Mesh1P` components are added in the **Components** view, whereas the `TurnRateGamepad` variable is added to the **Details** panel because the `TurnRateGamepad` variable is an attribute of the class and not a component.

MyShooterProjectile.h and MyShooterProjectile.cpp

The `AMyShooterProjectile` class is defined for instantiating bullets that can be fired during the gameplay. Bullets are spawned in front of gun muzzles and move forward, either until they hit something or until the end of their lifespan, so the class only needs two components: `ProjectileMovement` and `USphereComponent` (see *Figure 4.16*).

A Class C++ function can be declared as a **UFunction** (the `UFUNCTION` macro will be introduced in *Chapter 5*) so that the function can be recognized by the engine's Reflection System. In *Figure 4.16*, you may notice that the `OnHit` function is designated as a UFunction, which enables the engine to recognize and associate it with the `OnComponentHit` delegate event of `UsphereComponent`:

```
4
5   #include "CoreMinimal.h"
6   #include "GameFramework/Actor.h"
7   #include "MyShooterProjectile.generated.h"
8
9   class USphereComponent;
10  class UProjectileMovementComponent;
11
12  UCLASS(config=Game)
13  class AMyShooterProjectile : public AActor
14  {
15      GENERATED_BODY()
16
17      /** Sphere collision component */
18      UPROPERTY(VisibleDefaultsOnly, Category=Projectile)
19      USphereComponent* CollisionComp;
20
21      /** Projectile movement component */
22      UPROPERTY(VisibleAnywhere, BlueprintReadOnly, Category = Movement, meta = (AllowPrivateAccess = "true"))
23      UProjectileMovementComponent* ProjectileMovement;
24
25  public:
26      AMyShooterProjectile();
27
28      /** called when projectile hits something */
29      UFUNCTION()
30      void OnHit(UPrimitiveComponent* HitComp,
31          AActor* OtherActor, UPrimitiveComponent* OtherComp, FVector NormalImpulse, const FHitResult& Hit);
32
33      /** Returns CollisionComp subobject **/
34      USphereComponent* GetCollisionComp() const { return CollisionComp; }
35      /** Returns ProjectileMovement subobject **/
36      UProjectileMovementComponent* GetProjectileMovement() const { return ProjectileMovement; }
37  };
38
39
```

Figure 4.16 – MyShooterProjectile.h

Here, *lines 30-31* declare the `OnHit` function with the `UFUNCTION()` macro above them.

`MyShooterProjectile.cpp` only implements two functions – the `class` constructor and the `OnHit` function:

```
MyShooterProjectile.cpp*  ⊕ ×   MyShooterProjectile.h*     MyShooterCharacter.h     MyShooterCharacter.cpp
MyShooter                              ▼ ↓ AMyShooterProjectile          ▼ ◐ AMyShooterProjectile()
    3      #include "MyShooterProjectile.h"
    4      #include "GameFramework/ProjectileMovementComponent.h"
    5      #include "Components/SphereComponent.h"
    6
    7      AMyShooterProjectile::AMyShooterProjectile()
    8      {
    9          // Use a sphere as a simple collision representation
   10          CollisionComp = CreateDefaultSubobject<USphereComponent>(TEXT("SphereComp"));
   11          CollisionComp->InitSphereRadius(5.0f);
   12          CollisionComp->BodyInstance.SetCollisionProfileName("Projectile");
   13          CollisionComp->OnComponentHit.AddDynamic(this, &AMyShooterProjectile::OnHit);
   14                          // set up a notification for when this component hits something blocking
   15
   16          // Players can't walk on it
   17          CollisionComp->SetWalkableSlopeOverride(FWalkableSlopeOverride(WalkableSlope_Unwalkable, 0.f));
   18          CollisionComp->CanCharacterStepUpOn = ECB_No;
   19
   20          // Set as root component
   21          RootComponent = CollisionComp;
   22
   23          // Use a ProjectileMovementComponent to govern this projectile's movement
   24          ProjectileMovement = CreateDefaultSubobject<UProjectileMovementComponent>(TEXT("ProjectileComp"));
   25          ProjectileMovement->UpdatedComponent = CollisionComp;
   26          ProjectileMovement->InitialSpeed = 3000.f;
   27          ProjectileMovement->MaxSpeed = 3000.f;
   28          ProjectileMovement->bRotationFollowsVelocity = true;
   29          ProjectileMovement->bShouldBounce = true;
   30
   31          // Die after 3 seconds by default
   32          InitialLifeSpan = 3.0f;
   33      }
   34
   35      void AMyShooterProjectile::OnHit(UPrimitiveComponent* HitComp,
   36          AActor* OtherActor, UPrimitiveComponent* OtherComp, FVector NormalImpulse, const FHitResult& Hit)
   37      {
   38          // Only add impulse and destroy projectile if we hit a physics
   39          if ((OtherActor != nullptr) && (OtherActor != this) && (OtherComp != nullptr) && OtherComp->IsSimulatingPhysics())
   40          {
   41              OtherComp->AddImpulseAtLocation(GetVelocity() * 100.0f, GetActorLocation());
   42
   43              Destroy();
   44          }
   45      }
```

Figure 4.17 – MyShooterProjecttile.cpp

Let's break down the code:

- *Lines 10-18* create and initialize the sphere collision component.

- *Lines 24-29* create and initialize the projectile movement component.

- *Line 32* sets the lifespan for the projectile. If the bullet doesn't hit anything after the lifespan time, it is destroyed.

- The `if` statement on *line 39* checks whether the projectile hits a valid object.

- *Line 41* adds some force to the hit object.

- *Line 43* destroys the projectile after it hits the object.

The `MyShooterProjectile` class serves as an example that demonstrates how to control the movement of an actor within the game and effectively handle collisions. By studying this class, developers can gain insights into implementing projectile behavior and incorporating collision detection mechanisms into their own games.

TP_PickUpComponent.h and TP_PickUpComponent.cpp

The `UTP_PickUpComponent` class inherits from `USphereComponent` and adds the feature that handles the overlap event. When a player enters this sphere area, the weapon with this component is picked up.

The main task of the header file is to define an `OnPickUp` delegate, which will be called when the `OnSphereBeginOverlap` event is triggered:

```
TP_PickUpComponent.h* ⊕ ×   TP_PickUpComponent.cpp      MyShooterProjectile.cpp*      MyShooterProjectile.h*                « ⬦
MyShooter                              ⬦  UTP_PickUpComponent                   ⬦  OnSphereBeginOverlap(UPrimitiveComponent * ( ⬦
    3        #pragma once
    4
    5      ⊟#include "CoreMinimal.h"
    6       │#include "Components/SphereComponent.h"
    7       │#include "MyShooterCharacter.h"
    8       │#include "TP_PickUpComponent.generated.h"
    9
   10      ⊟// Declaration of the delegate that will be called when someone picks this up
   11       │// The character picking this up is the parameter sent with the notification
   12        DECLARE_DYNAMIC_MULTICAST_DELEGATE_OneParam(FOnPickUp, AMyShooterCharacter*, PickUpCharacter);
   13
   14        UCLASS(Blueprintable, BlueprintType, ClassGroup = (Custom), meta = (BlueprintSpawnableComponent))
   15      ⊟class MYSHOOTER_API UTP_PickUpComponent : public USphereComponent
   16       │{
   17            GENERATED_BODY()
   18
   19        public:
   20
   21            /** Delegate to whom anyone can subscribe to receive this event */
   22            UPROPERTY(BlueprintAssignable, Category = "Interaction")
   23            FOnPickUp OnPickUp;
   24
   25            UTP_PickUpComponent();
   26        protected:
   27
   28            /** Called when the game starts */
   29            virtual void BeginPlay() override;
   30
   31            /** Code for when something overlaps this component */
   32            UFUNCTION()
   33            void OnSphereBeginOverlap(UPrimitiveComponent* OverlappedComponent,
   34                AActor* OtherActor, UPrimitiveComponent* OtherComp, int32 OtherBodyIndex,
   35                bool bFromSweep, const FHitResult& SweepResult);
   36       │};
```

Figure 4.18 – TP_PickUpComponent.h

Let's break down the code:

- *Line 12* uses the `DECLARE_DYNAMIC_MULTICAST_DELEGATE_OneParam` macro to define the delegate, and this delegate requires the subscribing function to have a single parameter of type `AMyShooterCharacter*`.

- The UCLASS macro in *line 14* has some specifiers here:

  - `Blueprintable` indicates that this class is an acceptable base class for creating Blueprints.

  - `BlueprintType` indicates that this class can be used for variables in Blueprints.

  - `ClassGroup = (Custom)` indicates that this class should be within the `Custom` group.

  - `meta = (meta tags)` informs the engine to use the meta tags as instructed. For instance, the `BlueprintSpawnableComponent` tag tells the engine that this component can be spawned by Blueprint, and the class name will show on the list of available components for the Blueprint **SpawnActor** node. Another example is when you want to display a class with a different name from its original class name of `UTP_PickupComponent` in the editor; add `meta = (DisplayName="New Class Name")` as the specifier to the UCLASS macro.

- *Line 23* defines the `OnPickUp` delegate variable.

- The `UFUNCTION()` macro on *line 32* allows the C++ `OnSphereBeginOverlap` function to be recognized by the Unreal Engine Reflection System. This function must be a `UFUNCTION` to be added to the `OnComponentBeginOverlap` delegate in the base class.

Based on the declaration of the delegate and the event handler function in the header file, the constructor of the class and the `BeginPlay` and `OnSphereBeginOverlap` functions are implemented in `TP_PickupComponent.cpp`:

```
TP_PickUpComponent.cpp  ⊕ ×  TP_PickUpComponent.h      MyShooterProjectile.cpp      MyShooterProjectile.h
MyShooter                        ↓ UTP_PickUpComponent                        OnSphereBeginOverlap(UPrimitiveCompc
   3      #include "TP_PickUpComponent.h"
   4
   5      UTP_PickUpComponent::UTP_PickUpComponent()
   6      {
   7          // Setup the Sphere Collision
   8          SphereRadius = 32.f;
   9      }
  10
  11      void UTP_PickUpComponent::BeginPlay()
  12      {
  13          Super::BeginPlay();
  14
  15          // Register our Overlap Event
  16          OnComponentBeginOverlap.AddDynamic(this, &UTP_PickUpComponent::OnSphereBeginOverlap);
  17      }
  18
  19      void UTP_PickUpComponent::OnSphereBeginOverlap(UPrimitiveComponent* OverlappedComponent,
  20          AActor* OtherActor, UPrimitiveComponent* OtherComp, int32 OtherBodyIndex,
  21          bool bFromSweep, const FHitResult& SweepResult)
  22      {
  23          // Checking if it is a First Person Character overlapping
  24          AMyShooterCharacter* Character = Cast<AMyShooterCharacter>(OtherActor);
  25          if(Character != nullptr)
  26          {
  27              // Notify that the actor is being picked up
  28              OnPickUp.Broadcast(Character);
  29
  30              // Unregister from the Overlap Event so it is no longer triggered
  31              OnComponentBeginOverlap.RemoveAll(this);
  32          }
  33      }
  34
```

Figure 4.19 – TP_PickUpComponent.cpp

Let's break down the code:

- *Line 8* initializes `SphereRadius`.
- *Line 13* calls the `BeginPlay` function of the `USphereComponent` base class.
- *Line 16* adds the `OnSphereBeginOverlap` delegate function.
- *Line 24* calls the `Cast` template function, which casts the input `AActor` pointer to be a `MyShooterCharacter` pointer. If the cast fails, the return value is `nullptr`, which means it is a null pointer. In Unreal Engine, functions are usually generalized to take in parameters and return values in the base class pointers. When you know what an actual pointer type is, you can cast it so that you can access the right class members.
- *Line 25* checks whether the collided character is valid (`Character != nullptr`). When the condition is true, *lines 28-31* in the code are executed.

Please be aware that extending collision components and handling the `OnComponentBeginOverlap` event by associating event functions is very commonly used in processing game interactions.

TP_WeaponComponent.h and TP_WeaponComponent.cpp

The weapon component takes care of firing processes, including playing the firing animation, spawning a projectile, playing sound effects, and attaching the weapon to a character who has picked it up.

Specifically, the `TP_WeaponComponent.h` file defines properties and functions for the class:

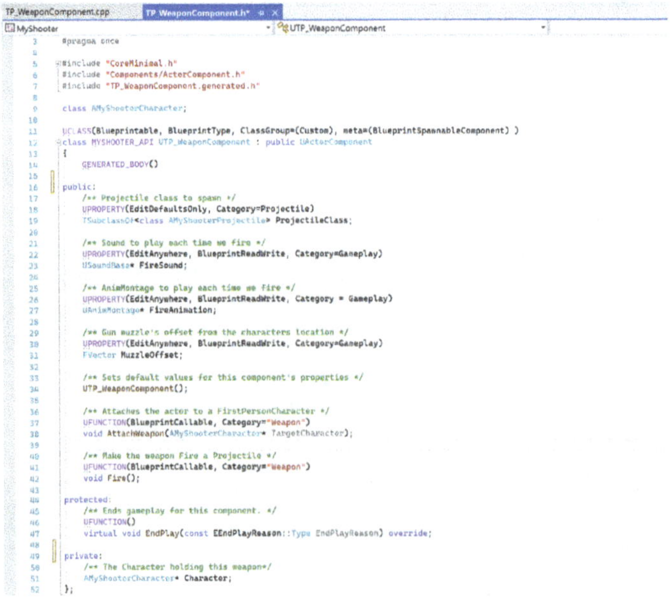

```
#pragma once

#include "CoreMinimal.h"
#include "Components/ActorComponent.h"
#include "TP_WeaponComponent.generated.h"

class AMyShooterCharacter;

UCLASS(Blueprintable, BlueprintType, ClassGroup=(Custom), meta=(BlueprintSpawnableComponent) )
class MYSHOOTER_API UTP_WeaponComponent : public UActorComponent
{
    GENERATED_BODY()

public:
    /** Projectile class to spawn */
    UPROPERTY(EditDefaultsOnly, Category=Projectile)
    TSubclassOf<class AMyShooterProjectile> ProjectileClass;

    /** Sound to play each time we fire */
    UPROPERTY(EditAnywhere, BlueprintReadWrite, Category=Gameplay)
    USoundBase* FireSound;

    /** AnimMontage to play each time we fire */
    UPROPERTY(EditAnywhere, BlueprintReadWrite, Category = Gameplay)
    UAnimMontage* FireAnimation;

    /** Gun muzzle's offset from the characters location */
    UPROPERTY(EditAnywhere, BlueprintReadWrite, Category=Gameplay)
    FVector MuzzleOffset;

    /** Sets default values for this component's properties */
    UTP_WeaponComponent();

    /** Attaches the actor to a FirstPersonCharacter */
    UFUNCTION(BlueprintCallable, Category="Weapon")
    void AttachWeapon(AMyShooterCharacter* TargetCharacter);

    /** Make the weapon Fire a Projectile */
    UFUNCTION(BlueprintCallable, Category="Weapon")
    void Fire();

protected:
    /** Ends gameplay for this component. */
    UFUNCTION()
    virtual void EndPlay(const EEndPlayReason::Type EndPlayReason) override;

private:
    /** The Character holding this weapon*/
    AMyShooterCharacter* Character;
};
```

Figure 4.20 – TP_WeaponComponent.h

Let's break down the code:

- *Lines 18-31* define the class properties.

- *Line 34* is the class constructor.

- *Lines 37-42* declare the two functions: AttachWeapon and Fire. Both these functions have the BlueprintCallable specifier, which indicates that they are visible to Blueprints.

- *Line 47* defines the EndPlay event function.

- *Line 51* defines an AMyShooterCharacter pointer, which points to the player character who picked the weapon up.

The TP_WeaponComponent.cpp file contains useful gameplay features for scripting implementations. From this part, you can learn how to spawn actors, play a sound at a location, play animations, and attach and detach delegate functions:

- **Spawning an actor**: Spawning characters frequently happens while playing a game. Shooting projectiles is a typical example of dynamic spawning. To spawn an actor, you need to get the pointer to the current game world first by calling the GetWorld() function:

  ```
  UWorld* const World = GetWorld();
  ```

  The const keyword between the (UWorld*) data type and the (World) variable name indicates that the retrieved object's contents are unchangeable.

  Now, you can call the world's SpawnActor function to actually create a new actor and place it in the world:

  ```
  World->SpawnActor<AMyShooterProjectile>(ProjectileClass,
  SpawnLocation, SpawnRotation, ActorSpawnParams);
  ```

  Here, the SpawnActor function has four parameters that tell the engine what type of actor is going to be spawned and where to place the new actor.

- **Playing a sound at a location**: To play the firing sound effect, the weapon's fire() function simply calls the PlaySoundAtLocation() function of the UGameplayStatics static class (a class that is never instantiated and provides static functions), which provides useful gameplay utility functions. The sound play function needs a location parameter because it plays stereo sound effects:

  ```
  UGameplayStatics::PlaySoundAtLocation(this, FireSound,
  Character->GetActorLocation());
  ```

  The function's first parameter, this, is a special alias in C++ that means the owner object itself. In this example, it means the weapon component.

- **Playing an animation**: To play the firing animation, it calls the `Montage_Play` function of the animation instance. This function takes two parameters – the animation and the speed of the animation:

```
AnimInstance->Montage_Play(FireAnimation, 1.f);
```

- **Attaching and detaching the Fire() function to the OnUseItem delegate variable defined in AMyShooterCharacter**: When the `AttachWeapon` function of the `UTP_WeaponComponent` class is called, it registers its `Fire` function pointer to the character's `OnUseItem` delegate variable:

```
Character->OnUseItem.AddDynamic(this, &UTP_
WeaponComponent::Fire);
```

When the `EndPlay` event function is called, the `Fire` function is unregistered from the character's `OnUseItem` delegate variable:

```
Character->OnUseItem.RemoveDynamic(this, &UTP_
WeaponComponent::Fire);
```

The engine APIs mentioned previously presented techniques for spawning actors, playing 3D sound effects, executing montage animations, and dynamically attaching and detaching delegate event functions. These methods are valuable references for your future game development.

MyShooter.h and MyShooter.cpp

The `MyShooter` module simply does the initial tasks for starting the game:

```
MyShooter.cpp  ⊕ ✕  MyShooter.h
⊞ MyShooter                                          ▼   (Global Scope)                                      ▼
     1        // Copyright Epic Games, Inc. All Rights Reserved.
     2
     3      ⊟#include "MyShooter.h"
     4       #include "Modules/ModuleManager.h"
     5
     6        IMPLEMENT_PRIMARY_GAME_MODULE( FDefaultGameModuleImpl, MyShooter, "MyShooter" );
     7
```

Figure 4.21 – MyShooter.cpp

The only line of code in `MyShooter.cpp` is the `IMPLEMENT_PRIMARY_GAME_MODULE` macro, which designates that the project's primary module is `MyShooter`. The root directory of the primary module is `Source/MyShooter`, where the `MyShooterBuild.cs` build file should reside.

MyShooterGameMode.h and MyShooterGameMode.cpp

Every Unreal project has a `GameMode` object that handles information about the game played. The `AMyShooterGameMode` class defined in this module extends the engine's `AGameMode` class but doesn't add anything new to it:

```
MyShooterGameMode.h  ⇌  ×
MyShooter                                    ▾  (Global Scope)                                    ▾
     1      // Copyright Epic Games, Inc. All Rights Reserved.
     2
     3      #pragma once
     4
     5    ⊟#include "CoreMinimal.h"
     6     │#include "GameFramework/GameModeBase.h"
     7     │#include "MyShooterGameMode.generated.h"
     8
     9      UCLASS(minimalapi)
    10    ⊟class AMyShooterGameMode : public AGameModeBase
    11     │{
    12     │    GENERATED_BODY()
    13
    14     │public:
    15     │    AMyShooterGameMode();
    16     │};
```

Figure 4.22 – MyShooterGameMode.h

MyShooterGameMode has no additional information and currently behaves identically to its base class. Keep in mind that AMyShooterGameMode is the place where you add new game rules, such as level transition and game-specific behaviors in future development.

MyShooter.Build.cs, MyShooter.Target.cs, and MyShooterEditor.target.cs

These C# files contain the build settings information. Only the build.cs file may need to be edited to add or remove modules for advanced programming needs. There should be no need to make any changes to the two target.cs files.

Here is the code from MyShooter.Build.cs:

```
MyShooter.Build.cs*  ⇌  ×
Miscellaneous Files                          ▾  ⚙MyShooter
     1      // Copyright Epic Games, Inc. All Rights Reserved.|
     2
     3      using UnrealBuildTool;
     4
     5    ⊟public class MyShooter : ModuleRules
     6     │{
     7    ⊟    public MyShooter(ReadOnlyTargetRules Target) : base(Target)
     8     │    {
     9     │        PCHUsage = PCHUsageMode.UseExplicitOrSharedPCHs;
    10     │
    11    ⊟        PublicDependencyModuleNames.AddRange(new string[] {
    12     │            "Core", "CoreUObject", "Engine", "InputCore", "HeadMountedDisplay"
    13     │        });
    14     │    }
    15     │}
    16
```

Figure 4.23 – MyShooter.build.cs

The MyShooter.build.cs file specifies that the project should include the five modules of *Core*, *CoreUObject*, *Engine*, *InputCore*, and *HeadMountedDisplay*. When building the project, the engine packs those modules into the final package so that the supported features are available in the game.

The two `target.cs` files are configurations for different build types – *Game* or *Editor* and *Client* or *Server*, for example. The two `target.cs` files are not our concern here, so we'll skip delving into the details of these files.

In this section, we have gone through the essential parts of the source code and the project configurations. As `MyShooter` is an Unreal C++ project, you have an additional option to open the engine editor and the project. Rather than going through the **Epic Games Launcher**, you can directly launch the engine and open the game project in Visual Studio. Let's explore this new option.

Launching Unreal Editor and opening the game project in Visual Studio

You already know how to open an existing Unreal C++ project in Unreal Editor through the Epic Games Launcher – this is the standard, easy method in most situations – but another way to launch Unreal Editor and open your game project is by directly running the program in Visual Studio.

The main benefits of starting Unreal Editor and opening game projects in Visual Studio are as follows:

- Debugging the source code and troubleshooting bugs
- Having accessibility to the engine source
- Customizing the engine for special needs
- Fixing engine bugs

Follow these steps to open the game project in VS:

1. Launch Visual Studio.
2. Open the `MyShooter.sln` C++ solution.
3. Choose the right build configuration from **Solution Configurations**:

Figure 4.24 – Visual Studio Solution Configurations

The build configuration options are as follows:

- **DebugGame** builds the engine code with optimizations and game code with debugging symbols

- **DebugGame Editor** does the same work as **DebugGame** and loads the engine editor

- **Development** builds the engine and the game code with optimizations for some time-consuming code and adds debugging symbols to the rest

- **Development Editor** does the same work as **Development** and also loads the engine editor

- **Shipping** builds the engine and game code with the most optimized performance for shipping products

4. Build the solution (if needed).

5. Start running the solution by clicking **Start Debugging** (the solid green play button) or **Start Without Debugging** (the play button with a green outline) on the toolbar. You can also find the menu items on the **Debug** menu:

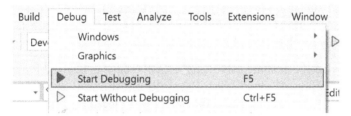

Figure 4.25 – Visual Studio Debug menu

For debugging purposes, it is highly recommended to utilize the **Development Editor** option. This configuration enables using a comprehensive set of debug features, such as toggling breakpoints, step-by-step tracing, and observing variable values, offering methods for efficient troubleshooting processes.

Please be aware that one important benefit of opening game projects in Visual Studio is that you can debug not only your own game source code but also the engine code. On the other hand, when closing the project, please do so in the editor rather than in Visual Studio. By doing this, you can avoid losing unsaved editing work.

Besides the convenience of code debugging, it is crucial to note that another significant advantage of opening game projects in Visual Studio is the ability to trace not only the game source code but also the engine code. However, when it comes to closing the project, again, it is recommended to do so within the editor rather than directly in Visual Studio. This practice helps prevent the loss of unsaved editing work, ensuring that any changes made are properly saved before the project is closed.

Summary

We just investigated the C++ MyShooter project by looking at the basic project structure and reviewing the source code. You should now have an overall idea about what the C++ scripts do and how they collaborate with the engine and the Blueprints.

From the sample game code, you have also learned some useful Unreal C++ scripting skills.

First, we reviewed the code for creating Unreal recognizable C++ classes by marking them with the UCLASS macro and the specifiers. You now understand that the Unreal Engine Reflection System will use the information associated with the macros and the specifiers to spawn objects and components. You then got to know some Unreal base classes such as ACharacter, AActor, USphereComponent, UActorComponent, and AGameModeBase that can be inherited to create new game classes.

Second, we looked at the code that defines class variables and declares class functions with the UPROPERTY and UFUNCTION macros. Like the UCLASS macro you learned about before, these macros can also have specifiers and work with the Unreal Reflection System. We also introduced the three basic Actor functions: BeginPlay(), EndPlay(), and Tick(). The DECLARE_ DYNAMIC_MULTICAST_DELEGATE_OneParam macro for declaring delegated function types and how to dynamically attach delegated functions were also explained.

Third, you also learned about some useful engine APIs that can be used for spawning actors (SpawnActor()), playing 3D sound effects (PlaySoundAtLocation()), playing montage animations (Montage_Play()), and getting the current actor's location (GetActorLocation()).

The last thing you learned was how to launch the Unreal Editor and open the C++ project in Visual Studio. This method avoids going back and forth between the **Epic Games Launcher** and the engine editor.

Based on the knowledge gained in this chapter, the upcoming chapter will go deeper into Unreal C++ scripting skills, including the creation of actor classes, defining UPROPERTYs and UFUNCTIONs, and more. To facilitate exploration and practical application, we will start a new top-down game named *Pangaea* that will provide hands-on experience and further insights into Unreal C++ development.

Part 2 –
C++ Scripting for
Unreal Engine

In this part, we will introduce the essential classes in Unreal Engine and the C++ scripting skills required to create the top-down game *Pangaea*. Key topics we will cover include actor creation, player input, character animation control, and game interactions.

Additionally, we will explore useful game-related engine features, such as spawn and despawn, input map settings, state machines, collision settings, navigation, and physics ray casting, that are required for developing the game.

The last chapter of this part delves into commonly employed software development processes for ensuring high-quality coding practices, offering insights into real-world development scenarios. By covering aspects of code refactoring and refinement, you will gain a better understanding of practical software development processes.

This part contains the following chapters:

- *Chapter 5, Learning How to Use the UE Gameplay Framework Base Classes*
- *Chapter 6, Creating Game Actors*
- *Chapter 7, Controlling Characters*
- *Chapter 8, Handling Collisions*
- *Chapter 9, Improving C++ Code Quality*

5

Learning How to Use UE Gameplay Framework Base Classes

Based on what you learned in the previous chapters, you should already know how to write C++ scripts for UE games and have an overall view of a typical UE C++ project structure. Now is the time to start learning basic C++ scripting skills in this chapter.

Games usually include a game environment, some actors (or game objects), and the interactions between the actors. Player characters are controlled by players, whereas **non-player characters** (**NPCs**) are controlled by game logic or **artificial intelligence** (**AI**) agents. In this chapter, you will learn how to derive base classes from Unreal Engine's gameplay framework to create your own game actors and characters. Here, three game project configuration classes—PlayerController, GameMode, and GameInstance—are going to be introduced, which will help us to define the game's specific players and rules.

Knowing the framework base classes is fundamental to writing C++ scripts for Unreal Engine. All the game actors and components are built up based on the engine's base classes.

The following topics will be covered in this chapter:

- Creating a *Pangaea* top-down game project
- Understanding the gameplay framework base classes
- Creating game actor classes
- Recompiling C++ projects
- Using the UPROPERTY macro
- Using the UFUNCTION macro
- Adding components to the new actors
- Creating blueprints from the new actor classes

- Learning about the Unreal gameplay framework classes
- Using the Cast template function

Technical requirements

The code for the Pangaea game project can be found at https://github.com/PacktPublishing/Unreal-Engine-5-Game-Development-with-C-Scripting/tree/main/Chapter05/Source.

The code for the My_CPP06 project can be found at https://github.com/PacktPublishing/Unreal-Engine-5-Game-Development-with-C-Scripting/tree/main/Chapter05/MyCPP_06.

Creating a Pangaea top-down game project

We believe that actually using what you have just learned to develop a real game is a very effective way for learners to quickly master new knowledge and skills.

Starting from this chapter, while still learning new C++ scripting skills, you will be working on a top-down game called Pangaea. The gameplay will involve controlling the main character running around the game map, killing enemies, and destroying enemy towers.

So, to get started, launch the UE5 editor from **Epic Games Launcher**:

Figure 5.1 – Steps to create the Pangaea project

Then perform the following steps to create the game project (see *Figure 5.1*):

1. Select **GAMES**.
2. Select **Top Down**.
3. Choose **C++**.
4. Choose the target directory (`C:\UEProjects`, for example).
5. Input the project name, which is `Pangaea`.
6. Then, click the **Create** button.

We have now created the game project. Let's have an overall view of the most used base classes as well as their relationships.

Understanding the gameplay framework base classes

Unreal Engine provides gameplay framework base classes for developers so that they can use and inherit from these base classes to create their new game-specific classes. Before utilizing the base classes, you need to understand the definitions of these classes as well as the relationships between them. The following class diagram will give you an overall view of the classes and the inheritance relationships:

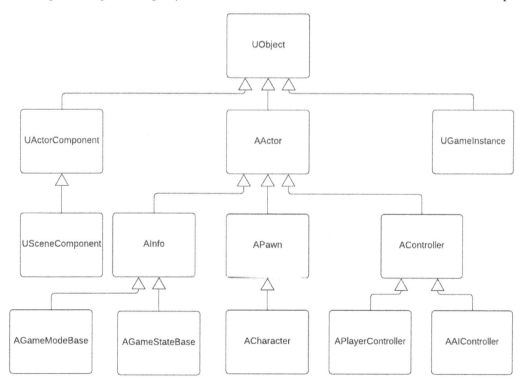

Figure 5.2 – UE5 gameplay framework classes diagram

From the diagram, you can see the following:

- UObject is the ancestor class of all the remaining classes
- AActor is the base class that is inherited by three groups of subclasses including the game actor (APawn), game information (AInfo), and player controller (AController) subclasses
- UActorComponent is the base class for all the component classes
- UGameInstance is a globally unique game instance manager class that can be extended to define game-specific variables, functions, and so on

Let's learn how to extend the aforementioned base classes to create some game-specific classes. When scripting, we will mainly follow the **UE5 coding standard** (please go to the official website https://docs.unrealengine.com/5.0/en-US/epic-cplusplus-coding-standard-for-unreal-engine for more details).

Creating game actor classes

The term *game actor classes* refers to the AActor, APawn, and ACharacter classes. These three classes are used to instantiate game actors that will be placed in the game levels, as follows:

- AActor is the base class for creating a wide range of objects, such as buildings, spawn points, portals, vehicles, characters, and so on. We will extend this class to create ADefenseTower, AWeapon, and AProjectile classes.
- APawn is a subclass of AActor that is used to create non-character, player-controllable actors (not characters) that accept and react to player inputs—racing cars, for example.
- ACharacter extends the APawn class for creating characters. A character can not only accept user inputs and moves but also has at least one skeletal mesh and the character state animations, such as *idle*, *walk*, *run*, *attack*, *die*, and so on. We will extend this class to create a new APlayerAvatar class.

Let's practice creating the three important gameplay actor classes, ADefenseTower (the defense tower), AProjectile (the projectile that defense towers shoot), and the player character (APlayerAvatar) for the Pangaea game.

Creating the ADefenseTower class

The ADefenseTower class will be used to place a defense tower on game maps. A defense tower fires projectiles to attack the player when the player character is within its attack range. Defense towers have their life points—when a tower is hit by the player, the life point value drops and the tower is destroyed when the life point value is lower or equal to 0.

Perform the following steps to create the `ADefenseTower` class:

1. Select `Pangaea│All│C++ Classes│Pangaea` from the **Content Browser**.

2. Right-click on an empty area in the source code browser and choose **New C++ Class...** from the pop-up menu or select **Tools** | **New C++ Class...** from the UE editor's main menu:

Figure 5.3 – Creating a new C++ class

3. Since defense towers don't need to accept player inputs, you can choose `AActor` to be the base class for creating the `ADefenseTower` class. Then, click the **Next>** button to continue:

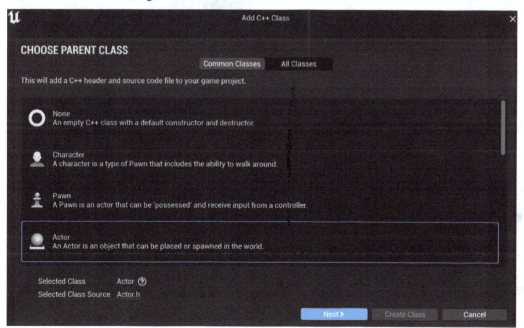

Figure 5.4 – Choosing a base class

4. Next, leave the class type's **Public** and **Private** buttons unselected. Then, type `DefenseTower` into the **Name** field, and click the **Create Class** button (note that Unreal will automatically add the prefix `A` to the new class name and make it `ADefenseTower`):

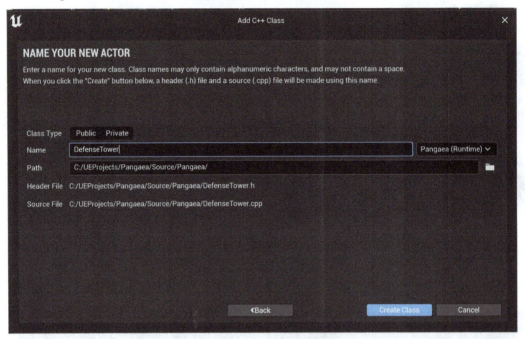

Figure 5.5 – Setting up the base class

After completing these steps, the `DefenseTower` class should show up in the class list, like so:

Figure 5.6 – The DefenseTower class has been created and now shows in the class list

5. Now, double-click to open the class's `DefenseTower.cpp` source code file in Visual Studio. Then, open `DefenseTower.h`, where you will find the following code in the header file:

```
#pragma once
#include "CoreMinimal.h"
#include "GameFramework/Actor.h"
#include "DefenseTower.generated.h"

UCLASS()
class PANGAEA_API ADefenseTower : public AActor
{
        GENERATED_BODY()
protected:
        virtual void BeginPlay() override;
public:
        virtual void Tick(float DeltaTime) override;
};
```

6. Here, we can add some basic class attributes (variables) and member functions to the `ATowerDefence` class. First, add the **public attributes**, like so:

```
int HealthPoints = 100;
int ShellDefense = 2;
float AttackRange = 15.0f;
float ReloadInterval = 1.0f;
```

Public attributes usually have fixed values that are not frequently changed during runtime, but these values should be configurable either during editing time or set at the beginning of gameplays. For example, a tower's `AttackRange` value is set to 15, which means that when the player character is within this range, the tower fires at it. This value may be increased only when the tower levels up or is purposely set with a different value in the editor.

7. Then, add the **protected attributes**, as follows:

```
int _HealthPoints;
float _ReloadCountingDown;
```

Protected attributes can be accessed within the class itself or its child classes. Their values are changed during gameplay time to indicate some gameplay states. For example, `ReloadCountingDown` is set to be the value of `ReloadInterval`, and its value will be reduced every tick. The tower cannot do the next fire until `_ReloadCountingDown` goes back to 0.

You may have noticed that we added an underscore (_) prefix for the protected variables. It is recommended to name private and protected variables this way to distinguish them from public variables.

8. Next, add the **public functions**, like so:

```
int GetHealthPoints();
bool IsDestroyed();
bool CanFire();
void Fire();
void Hit(int damage);
```

Public functions can be called either internally or by their child classes and outside callers. To determine whether a function is public, protected, or private, you should consider the actual requirements and designs.

9. Finally, add the **protected functions**, as follows:

```
void DestroyProcess();
```

Unlike private functions that only can be called internally, in addition, protected functions also can be called by their child classes. In this case, the DestroyProcess() function is called when the tower's life point reaches 0.

The ADefenseTower header file code should now look like this:

```
#pragma once
#include "CoreMinimal.h"
#include "GameFramework/Actor.h"
#include "DefenseTower.generated.h"

UCLASS()
class PANGAEA_API ADefenseTower : public AActor
{
        GENERATED_BODY()
public:
        ADefenseTower();
        int HealthPoints = 100;
        int ShellDefense = 2;
        float AttackRange = 15.0f;
        float ReloadInterval = 1.0f;
protected:
        virtual void BeginPlay() override;

        int _HealthPoints; //the tower's current health points
        float _ReloadCountingDown;
public:
        virtual void Tick(float DeltaTime) override;

        int GetHealthPoints();
        bool IsDestroyed();
```

```
        bool CanFire();
        void Fire();
        void Hit(int damage);
protected:
        void DestroyProcess();
};
```

Now, let's move on to the `AProjectileTower` class.

Creating the AProjectile class

Projectiles can be fired out by defense towers. Once a projectile is fired, it flies along its initial direction at a constant speed. If the projectile hits something, it deals damage to the hit object and destroys itself; otherwise, if the projectile runs out of its lifetime, it is destroyed too.

The `AProjectile` class can be inherited as child classes for creating various fireable objects, such as `AFireBall`, `AMissile`, `ABomb`, and so on. We will work on creating such child classes in the future.

You can follow the steps you learned for creating the `ADefenseTower` class to create the `AProjectile` class and add the following attributes to the class:

1. Add the public attributes, like so:

    ```
    float Speed = 100.0f;
    float Lifespan = 5.0f;
    float Damage = 10.0f;
    ```

2. Then, add the protected attribute, as follows:

    ```
    float _LifeCountingDown;
    ```

And that's it! Now, let's move on to the `APlayerAvatar` class.

Creating the APlayerAvatar class

Instead of the default pawn class, the `APlayerAvatar` class is going to be used by the engine to spawn the player character. The player character not only accepts and reacts to player inputs by playing character animations and moving the character but also handles events to simulate interactions among actors.

You can now choose **Character** as the base class for creating the `APlayerAvatar` class (see *Figure 5.7*):

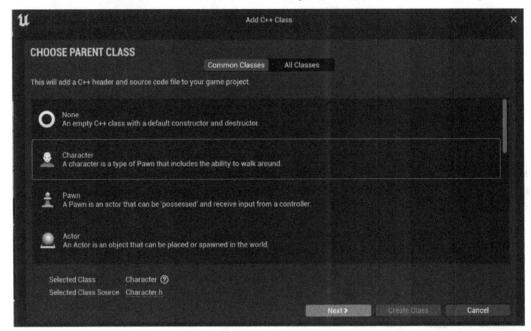

Figure 5.7 – Selecting Character as the base class for creating the APlayerAvatar class

---

Note

The main difference between `ACharacter` and `AActor` and `APawn` is that `ACharacter` contains the `SkeletalMesh`, `Movement`, and `Capsule` collider components by default, as well as the animation supports.

Now, add the following basic attributes to the `APlayerAvatar` class:

1. Add the public attributes, as follows:

```
int HealthPoints = 500;
float Strength = 10;
float Armer = 3;
float AttackRange = 6.0f;
float AttackInterval = 1.2f;
```

2. Add the protected attributes, like so:

```
int _HealthPoints;
float _AttackCountingDown;
```

3. Add the public functions, as follows:

```
int GetHealthPoints();
bool IsKilled();
bool CanAttack();
void Attack();
bool IsAttacking();
void Hit(int damage);
```

4. Add the protected functions, like so:

```
void DieProcess();
```

We have now added several new C++ classes derived from `AActor` and `ACharacter`, such as `ADefenseTower`, `AProjectile`, and `APlayerAvatar`, into the `Pangaea` project. However, adding new classes to an Unreal C++ project and making modifications to the source code necessitates recompiling the C++ project in Visual Studio. Let's now learn how to accomplish this work.

Recompiling C++ projects

Whenever you make changes to your C++ project, including code changes, adding new source files, and removing unused source files, you need to recompile the C++ project.

The simplest and most straightforward way is to click the **Recompile and Reload** button in the bottom-right corner of the UE editor:

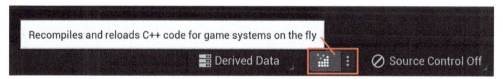

Figure 5.8 – The Recompile and Reload button recompiles and
reloads C++ code for game systems on the fly

Sometimes, this may not work because of the addition or removal of classes from the project. If this is the case, you can close the UE editor and build the project or the solution in Visual Studio.

If you manually delete source files in **File Explorer**, you should regenerate the Visual Studio project files before recompiling. To complete this task, find your `Pangaea.uproject` file in **File Explorer**, right-click on the `.uproject` file, and choose **Generate Visual Studio project files** from the pop-up menu:

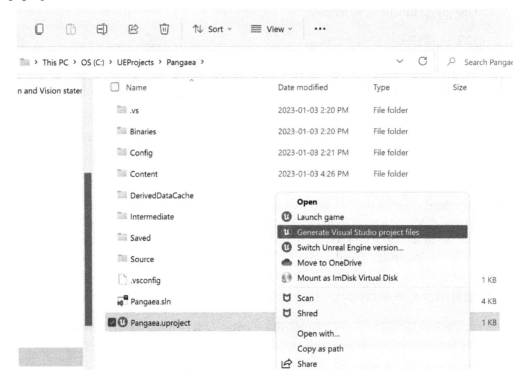

Figure 5.9 – Generating Visual Studio project files

Once the Visual Studio project files have been regenerated, our new C++ classes are seamlessly incorporated into the project. In the subsequent step, what we aim to explore involves leveraging Unreal's UPROPERTY and UFUNCTION macros to declare properties and functions that can be recognized and utilized by the engine.

Using the UPROPERTY macro

The UPROPERTY macro is placed above the definition of standard C++ class variables to declare Unreal-recognized class properties. The UPROPERTY macro can have specifiers and metadata for different use cases.

The UPROPERTY syntax

Let's take a look at the UPROPERTY syntax:

```
UPROPERTY([specifier1, specifier2, …], [meta(key1=value, key2=value2,
… )]
Type VariableName;
```

Let's break it down:

- As with function parameters, the specifiers and metadata are enclosed by a pair of parentheses
- The square brackets are used to indicate that the enclosed content is optional
- The ellipsis means that you can include more items
- The metadata keys are only valid in the editor, and not for any game logic

Let's look at two examples. The first example shows how to define a simple UPROPERTY variable:

```
UPROPERTY()
bool bHasWeapon;
```

This example defines the bHasWeapon property without any specifiers and metadata.

The second example shows a more complex UPROPERTY variable with specifiers:

```
UPROPERTY(EditAnywhere, Category=Params, Meta=(DisplayName="SPD"))
float Speed;
```

This example defines the Speed property. Let's look at this in a bit more detail:

- The EditAnywhere specifier indicates that this property can be edited in the editor's property window
- The Category specifier groups the property into the Params category name in the **Blueprint Editor** window
- The MetaData specifier contains only one key, which indicates that the property name will be displayed as **SPD** instead of **Speed** in the **Blueprint Editor** window

Having learned how to declare class variables as Unreal-recognized UPROPERTY macros, let us delve deeper into using UPROPERTY specifiers and metadata keys.

The UPROPERTY specifiers and metadata keys

In the following table, we will introduce you to just some of the UPROPERTY specifiers and metadata keys that are most used in this book. For details about UPROPERTY specifiers and metadata keys, you can visit the UE5 properties website (https://docs.unrealengine.com/5.0/en-US/unreal-engine-uproperties/):

Specifiers	
BlueprintAssignable	This delegate property can be assigned with a custom event function in Blueprint.
BlueprintAuthorityOnly	This is a delegate property that only accepts events with the BlueprintAuthorityOnly tag.
BlueprintReadOnly	This property is read-only in Blueprint.
BlueprintReadWrite	This property can be read and written in Blueprint.
Category = "name1|name2..."	This property belongs to a category in the Blueprint Editor. You can use | as the delimiter to define nested categories.
EditAnywhere	This property is editable anywhere in the editor.
VisibleAnywhere	This property is visible but read-only anywhere in the editor.
Metadata keys	
AllowPrivateAccess	This private property is accessible to Blueprint.
DisplayName = "Property Name"	This property should display the given Property Name value instead of the code-generated name in the editor.

Figure 5.10 – UPROPERTY specifiers and metadata keys

Now, we can mark the existing class variables of the actor classes to be UE properties so that we can configure the actor's properties in the editor.

Marking the ADefenseTower, AProjectile, and APlayerAvatar attributes as UE properties

Having learned about the usage of UPROPERTY and UFUNCTION, now is the opportunity to use the macros to designate our C++ class attribute variables as Unreal-recognized properties.

Open the header files and tag the attributes with the UPROPERTY macro. So, in ADefenseTower.h, enter the following code:

```
UPROPERTY(EditAnywhere, Category="Tower Params")
int HealthPoints = 500;
```

```
UPROPERTY(EditAnywhere, Category="Tower Params")
int ShellDefence = 3;
UPROPERTY(EditAnywhere, Category="Tower Params")
float AttackRange = 6.0f;

UPROPERTY(EditAnywhere, Category="Tower Params")
float ReloadInterval = 1.0f;
```

Then, in `AProjectile.h`, enter the following code:

```
UPROPERTY(EditAnywhere, Category = "Projectile Params")
float Speed = 100.0f;

UPROPERTY(EditAnywhere, Category = "Projectile Params")
float Lifespan = 5.0f;

UPROPERTY(EditAnywhere, Category = "Projectile Params")
float Damage = 10.0f;
```

Then, in `APlayerAvatar.h`, enter the following code:

```
UPROPERTY(EditAnywhere, Category = "PlayerAvatar Params")
int HealthPoints = 500;

UPROPERTY(EditAnywhere, Category = "PlayerAvatar Params")
float Strength = 10.0f;

UPROPERTY(EditAnywhere, Category = "PlayerAvatar Params")
float Armor = 3.0f;

UPROPERTY(EditAnywhere, Category = "PlayerAvatar Params")
float AttackRange = 6.0f;

UPROPERTY(EditAnywhere, Category = "PlayerAvatar Params")
float AttackInterval = 1.2f;
```

In the previous code snippets, we utilized the UPROPERTY macro to designate a set of properties. Additionally, we employed the EditAnywhere specifier to ensure that all these properties are visible and editable within the Unreal editor. Furthermore, the Category specifier was applied to group these properties in the Unreal editor for organized display.

As with tagging UE properties, Unreal also provides another macro for marking C++ functions to be UE-recognized functions. Let's learn about the UFUNCTION macro next.

Using the UFUNCTION macro

The UFUNCTION macro can be placed above the line of standard C++ function declarations. As with the UPROPERTY macro, UFUNCTION also has its specifiers and metadata to interpret the use of the functions.

The UFUNCTION syntax

Let's first check out the UFUNCTION syntax:

```
UFUNCTION([specifier1, specifier2, …], [meta(key1=value,
    key2=value2, … )]
ReturnType FunctionName([param1, param2, …]) [const];
```

Let's break it down:

- The square brackets are used to indicate that the enclosed content is optional
- The ellipsis means that you can include more items
- The metadata keys are only valid in the editor, not for any game logic

The following example demonstrates how to mark the GetHealthPoints() function to be a UFUNCTION macro with a displayname value of **Get HP**:

```
UFUNCTION(BlueprintCallable, Category="Player Avatar",
    Meta=(DisplayName="Get HP"))
int GetHealthPoints();
```

This example declares the GetHealthPoints() function. Let's look at this in a bit more detail:

- The BlueprintCallable specifier indicates that this function can be called by Blueprint
- The Category specifier groups the function into the **Player Avatar** category name in the Blueprint editing tools
- The MetaData specifier contains only one key, which indicates that the function node name will be displayed as **Get HP** instead of **GetHealthPoint** in the Blueprint graph editing window.

Having learned how to declare class functions as Unreal-recognized UFUNCTION macros, let us delve deeper into using UFUNCTION specifiers and metadata keys.

UFUNCTION specifiers and metadata keys

The following table introduces just some of the UPROPERTY specifiers and metadata keys that are most used in this book. For details about UPROPERTY specifiers and metadata keys, you can visit the UE5 UFunctions (`https://docs.unrealengine.com/5.0/en-US/ufunctions-in-unreal-engine/`) website:

Specifiers	
`BlueprintCallable`	This function can be a node and called in Blueprint.
`BlueprintPure`	This function doesn't change any owning object's data and generates a Blueprint node without the execution pin.
`BlueprintImplementableEvent`	This function can be implemented in Blueprint.
`BluprintNativeEvent`	This function's C++ implementation can be overridden by a Blueprint function.
`Category = "name1|name2..."`	This function belongs to a category in the Blueprint editor. You can use | as the delimiter to define nested categories.
Metadata keys	
`DisplayName = "Property Name"`	This property should display the given `Property Name` value instead of the code-generated name in the editor.

Figure 5.11 – UFUNCTION specifiers and metadata keys

Now, we can mark the existing class functions of the created actor classes to be UE functions so that they can be recognized by the engine as well as be called by blueprints.

Tagging ADefenseTower and APlayerAvatar member functions as UFUNCTION macros

Since we intend to handle most game logic on the C++ side rather than in Blueprint, only the getter functions are marked as UFUNCTION macros. The `AProjectile` class doesn't have any function that needs to be tagged.

Open the `PlayerAvatar.h` and `DefenseTower.h` header files and tag the existing functions with the UFUNCTION macro. Afterward, proceed to implement these functions in the `PlayerAvatar.cpp` and `DefenseTower.cpp` files.

Let's begin by applying the UFUNCTION macro to the functions in `DefenseTower.h`:

```
UFUNCTION(BlueprintCallable,
Category = "Pangaea|Defense Tower",
meta=(DisplayName="GetHP"))
int GetHealthPoints();
```

```
UFUNCTION(BlueprintCallable,
Category = "Pangaea|Defense Tower")
bool IsDestroyed();

UFUNCTION(BlueprintCallable,
Category = "Pangaea|Defense Tower")
bool CanFire();
```

This code uses the BlueprintCallable specifier for the getter functions. These getter functions can also be marked with the BlueprintPure specifier.

The following is another version of the code using the BlueprintPure specifier instead of the BlueprintCallable specifier:

```
UFUNCTION(BlueprintPure,
Category = "Pangaea|Defense Tower",
meta=(DisplayName="GetHP"))
int GetHealthPoints();

UFUNCTION(BlueprintPure,
Category = "Pangaea|Defense Tower")
bool IsDestroyed();

UFUNCTION(BlueprintPure,
Category = "Pangaea|Defense Tower")
bool CanFire();
```

In order to avoid compiling errors, now add the following function implementations to the ADefenseTower.cpp file:

```
int ADefenseTower::GetHealthPoints()
{
    return _HealthPoints;
}

bool ADefenseTower::IsDestroyed()
{
    return (_HealthPoints > 0.0f);
}
```

```
bool ADefenseTower::CanFire()
{
    return (_ReloadCountingDown <= 0.0f);
}
```

Having converted DefenseTower functions into UFUNCTION macros, we can now proceed to use the UFUNCTION macro to annotate the functions in PlayerAvatar.h, like so:

```
UFUNCTION(BlueprintCallable,
Category="Pangaea|PlayerCharacter",
meta=(DisplayName="Get HP"))
int GetHealthPoints();

UFUNCTION(BlueprintCallable,
Category = "Pangaea|PlayerCharacter")
bool IsKilled();

UFUNCTION(BlueprintCallable,
Category = "Pangaea|PlayerCharacter")
bool CanAttack();
```

To avoid compiling errors, now add the following function implementations to the APlayerAvatar.cpp file:

```
int APlayerAvatar::GetHealthPoints()
{
    return _HealthPoints;
}


bool APlayerAvatar::IsKilled()
{
    return (_HealthPoints <= 0.0f);
}

bool APlayerAvatar::CanAttack()
{
    return (_AttackCountingDown <= 0.0f);
}
```

We have tagged and implemented UFUNCTION macros for both the ADefenseTower and APlayerAvatar classes. Based on that, the last thing we want to do to complete our new C++ actor classes is to add some components to them.

Adding components to the new actors

Unreal provides useful components that can be added to actors; for example, you can add a static mesh component to an actor so that the actor is visually represented in the game levels. In our case, we want to add a box collision component and a static mesh component to `ADefenseTower` and `AProjectile`, as follows:

- `UBoxComponent`: This is added as the actor's root component for collision detection
- `UstaticMeshComponent`: This is added as the child of the actor's root component, which allows us to select and display a 3D mesh in the game levels

To use these two components, you should include their header files at the beginning of `ADefenseTower.h` and `AProjectile.h` files.

Including component header files

The order of `#include` statements in C++ is not a big concern, but you should ensure that the `*.generated.h` statement is placed as the last `include` statement and right before the `UCLASS` definition.

Here is the example from `ADefenseTower.h`:

```
#include "CoreMinimal.h"
#include "GameFramework/Actor.h"
#include "Components/BoxComponent.h"
#include "Components/StaticMeshComponent.h"
#include "DefenseTower.generated.h"

UCLASS()
Class ADefenseTower
```

Remember to add the two bold text lines to `AProjectile.h` file as well.

Now, we can define the pointer variables—`UBoxComponent* _BoxComponent`, for instance—of the component types to store the created component instances.

Defining private properties for these two components

To reference the two added components, we can define two pointer variables: `_BoxComponent` and `_MeshComponent`. These variables will allow us to interact with and manipulate the respective components as needed.

Open `DefenseTower.h` and add the following code:

```
private:
UPROPERTY(VisibleAnywhere, BlueprintReadOnly,
Category = "Tower Component",
meta = (AllowPrivateAccess = "true"))
UBoxComponent* _BoxComponent;

UPROPERTY(VisibleAnywhere, BlueprintReadOnly,
Category = "Tower Component",
meta = (AllowPrivateAccess = "true"))
UStaticMeshComponent* _MeshComponent;
```

Here, we used the `AllowPrivateAccess` meta tag to inform the engine that this private variable can be accessed in the editor.

Adding public getter functions to the components

Now, we can add the getter functions as well so that the components can be accessed outside of the class. In `DefenseTower.h`, we can add the following code:

```
Public:
FORCEINLINE UBoxComponent* GetBoxComponent() const
{
return _BoxComponent;
}
FORCEINLINE UStaticMeshComponent* GetMeshComponent() const
{
return _MeshComponent;
}
```

Let's quickly understand this code:

- `FORCEINLINE` is a UE macro that forces the compiler to insert a copy of the code instead of calling the function by its address. Small and frequently called functions are preferred to declaring them as inline functions to reduce the function call overhead.

- Placing the `const` keyword after a function declaration guarantees that the function doesn't modify its parameters or mutable class members via pointers or references internally.

Please be aware that components of a class have to be instantiated in the constructor of the class, so we will create `BoxComponent` and `StaticMeshComponent` components in the `ADefenseTower()` constructor of the `ADefenseTower` class.

Creating components in the class constructor

Open `DefenseTower.cpp` and add the following bold text code lines:

```
ADefenseTower::ADefenseTower()
{
PrimaryActorTick.bCanEverTick = true;

_BoxComponent = CreateDefaultSubobject<UBoxComponent>(
  TEXT("Box Collision"));
SetRootComponent(_BoxComponent);

_MeshComponent =CreateDefaultSubobject
  <UStaticMeshComponent>(TEXT("Static Mesh"));
_MeshComponent->SetupAttachment(_BoxComponent);
}
```

Let's break down the code:

- Setting `PrimaryActorTick.bCanEverTick` to `true` tells the engine that the `Tick()` function of this class needs to be called every frame or every interval (when the `TickInterval` value is greater than 0).

- The `CreateDefaultSubobject<Class>(ObjectName)` function instantiates an object of the bracketed `Class` type. The new object is managed by the engine's garbage collection manager, so you don't need to manually release the memory in the future.

- The `SetRootComponent` function sets the box collision component to be the actor's root component.

- Calling a component's `SetupAttachment(ParentComponent)` function attaches this component as a child of `ParentComponent`.

From the example provided, you have acquired knowledge on how to add components to an actor. Once we have completed all the necessary work on a C++ class, we can proceed to create Blueprint classes based on our newly created C++ actor classes. This will allow for further customization and implementation of the actor's behavior and functionality within Unreal Engine.

Creating blueprints by extending the actor classes

You may wonder why we would want to create Blueprint classes even after having C++ actor classes. The answer is in finding the right balance.

While it is totally acceptable to create actor classes solely with C++, particularly to increase performance, there is a trade-off involved—by exclusively relying on C++, you sacrifice the flexibility provided by blueprints. On the other hand, utilizing blueprints to handle relatively simple and editable data settings proves to be a more advantageous choice. It allows for easier tweaking and iteration without requiring code changes.

Thus, combining the power of both C++ and blueprints enables us to strike the ideal balance between performance, complexity, and flexibility in game development.

For instance, a character class for the initial settings and display meshes for different characters may vary; in such a scenario, it would be much easier and less work to edit them in the Blueprint editor rather than defining extra C++ subclasses and recompiling the project.

To create a Blueprint class from a C++ class, the C++ class must be tagged with the `Blueprintable` specifier. So, add the following specifier for all three classes' `UCLASS` macro:

```
UCLASS(Blueprintable)
```

Then, rebuild the project, re-launch the editor, and carry out the following steps:

1. Choose the appropriate folder where you want to create the new blueprint (for example, **All | Content | TopDown | Blueprints**) and right-click. From the pop-up menu, choose **Blueprint Class …**.

2. Search for the base class by typing the class name—`DefenseTower`, for example—in the **ALL CLASSES** window (see *Figure 5.12*).

3. Select the required class from the class tree and click **Select**:

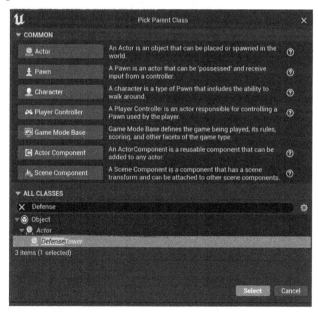

Figure 5.12 – Selecting a base class to create a new Blueprint

4. Rename the new Blueprint class with the format of a `BP_` prefix and the base class name— `BP_DefenseTower`, for example.

You can follow the previous steps to create blueprints for `BP_DefenseTower`, `BP_Projectile`, and `BP_PlayerAvatar`.

Once done, double-click `BP_ DefensePower` to open it, and have a close look in the Blueprint editor:

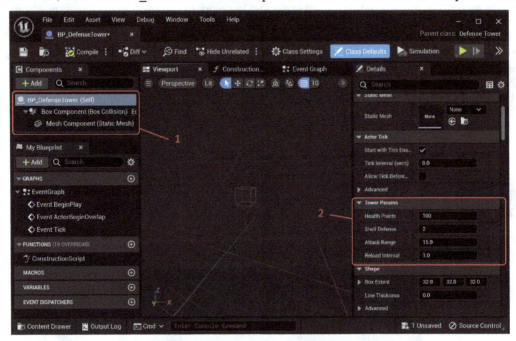

Figure 5.13 – BP_DefenseTower in the editor

Taking a look at the editor, we can see the following:

- *1* shows the **Components** view. It shows that the `Box` component is the root, and the `Mesh` component is the child of the `Box` component.

- *2* shows the **Details** panel where the properties under the **Tower Params** group are shown.

Now that we have learned how our new actor properties are displayed in the editor, we then want to explore how we can utilize our new actor `UFUNCTION` nodes in the event graph.

To find and use the `UFUNCTION` nodes on the event graph, navigate to the **Event Graph** window, right-click to open the **All Actions for this Blueprint** window, then find the `Pangaea` group, and expand it. Here, you should find the `DefenseTower` node, which contains three functions: `Can Fire`, `GetHP`, and `Is Destroyed`:

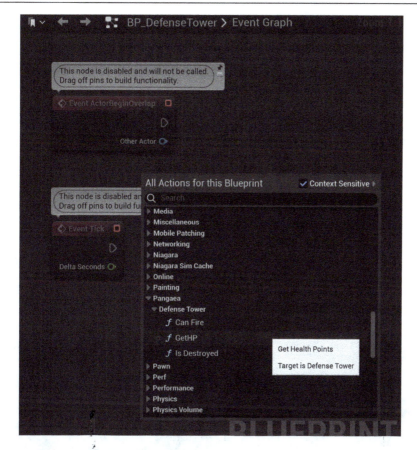

Figure 5.14 – Adding DefenseTower's GetHP function node onto the graph

When you select GetHP on the list of actions, the corresponding **Blueprint** node will be added and displayed in the **Event Graph** window in the editor.

Let's recall that we introduced the two versions of the GetHP function, utilizing either the BlueprintCallable or the BlueprintPure specifier. The two function versions will show the node slightly differently, as here:

Figure 5.15 – DefenseTower's GetHP function nodes

In the left node, the C++ function is tagged with the `BlueprintCallable` specifier, while in the right node, the C++ function is tagged with the `BlueprintPure` specifier.

Through the creation of the three actor classes for the `Pangaea` game, you have acquired knowledge on how to create actors with both C++ and Blueprint within Unreal Engine. Moving forward, let's learn about the four crucial gameplay framework classes.

Learning about the Unreal gameplay framework classes

The Unreal gameplay framework includes four classes, `PlayerController`, `GameModeBase`, `GameState`, and `GameInstance`, playing vital roles in managing various aspects of gameplay and overall game functionality. The first two classes are usually created automatically when a project is generated, while the remaining classes need to be manually added to the project.

Locating and creating gameplay framework classes in Pangaea

If you go to the **All** | **C++ Classes** | **Pangaea** folder, you should find that the project already has `PangaeaGameMode` and `PangaeaPlayerController` classes (see *Figure 5.16*). These two classes were created when the game project was initiated:

Figure 5.16 – The existing PangaeaGameMode and PangaeaPlayerController C++ classes

`PangaeaPlayerController` inherits from `PlayerController`, and `PangaeaGameMode` inherits from `GameModeBase`.

In addition to the existing `PangaeaGameMode` and `PangaeaPlayerController` classes, we can create `PangaeaGameState` and `PangaeaGameInstance` classes based on the following guidance:

1. Choose `GameStateBase` as the base class to create `PangaeaGameState`.

2. Choose `GameInstance` as the base class to create `PangaeaGameInstance`.

Now, the **C++ Classes | Pangaea** container has two more classes, as we can see here:

Figure 5.17 – Adding the PangaeaGameState and PangaeaGameInstance C++ classes

In the `Pangaea` project, we now have the gameplay framework classes in place. The next is to know the responsibilities of these classes and what they are used for. We will begin our exploration with the `PlayerController` class.

Learning about the PlayerController class

`PlayerController` is the engine's base C++ class, which can be extended for specific gameplay controls. `PangaeaPlayerController` is an instance that inherits from `PlayerController`, which encapsulates commonly used variables and functions needed for controlling player pawns.

It is not mandatory but recommended to consider using `PlayerController` in your game. `PlayerController` can be considered an invisible pawn that bridges the player and the controlled pawn or character. `PlayerController` decouples the player pawn's logic and view and makes it easier to let the player possess different pawns or characters.

When using `PlayerController`, you should consider letting it take care of the following responsibilities:

*   Receiving and handling player inputs

*   Moving and rotating the controlled pawn or character

*   Changing the state of the controlled pawn or character

*   Manipulating the camera views

You have discovered that `PlayerController` plays a vital role in managing interactions between players and the game. Now, let's delve into the `GameModeBase` class, which is the next class we are interested in.

Learning about the GameModeBase class

The `GameMode` object is needed for all UE projects. It stores important game information and settings, such as the starting level, the default player pawn, the player controller, and so on. By extending the `GameModeBase` class, you can add more gameplay information and rules to the child class. It could include the following:

- Flags that allow pausing the game, playing cinematics, enabling tutorials, and so on
- Level-transition conditions and processes
- Spawn locations
- For multiplayer games, the minimum number of players to start the game
- The conditions to end a game—the maximum game time, for example

For online games, the `GameMode` object only exists on the server side; we will discuss this topic later in *Chapter 11*.

GameState

The `GameState` object is usually used to store dynamic gameplay information. By extending the `GameStateBase` class, you can store important gameplay information in the child class. It could include the following:

- The time elapsed since the start of the game
- Total scores
- Waves of enemies
- Enemy positions for the minimap
- A countdown timer

For online games, the `GameState` object exists on both the server and client sides. The `GameState` object's information can be synced from the server to the clients.

GameInstance

The `GameInstance` object is a high-level manager representing the instance of a running game. By extending the `UGameInstance` class, you can put important local gameplay variables into the child class. In the context of online games, it's important to note that the `GameInstance` object exclusively exists on the client side.

Unreal games are internally built up on the multiplayer game framework, even if they may be single-player games. Therefore, during game development, you should always assume that the game was an online game and determine whether those global variables are necessary on the server. If a variable doesn't need to exist on the server, it can be considered a local gameplay variable and appropriately assigned as an attribute of the GameInstance class.

Obviously, the gameplay framework class instances encompass global variables that we may need to access from any part of our code. Thus, our next objective is to learn how to retrieve the instances to gain access to them.

Retrieving class instances from your code

To retrieve the previously introduced gameplay framework class instances from your code, you can refer to the following example:

```
UWorld* World = GetWorld();
APlayerController* PlayerController = World->
    GetFirstPlayerController();
AGameModeBase* GameMode = World->GetAuthGameMode();
AGameStateBase* GameState = World->GetGameState();
UGameInstance* GameInstance = World->GetGameInstance();
```

Observing the preceding sample code provided, you can discover that obtaining the game world instance is a prerequisite for acquiring instances of the gameplay framework classes. Once the world pointer is obtained, we can invoke its method functions to retrieve any of the framework class instances.

It is also noticeable from the previous sample code that all the retrieved variables are of the base class-type pointers instead of the desired child class-type pointers. This is because the engine only deals with generalized classes rather than our game-specific classes. However, it is possible to transform base class pointers into child class pointers by utilizing the powerful Cast function.

Using the Cast template function

Unreal Engine is designed as a universal graphics development tool. It provides base classes that can be inherited for developing specific games. That means the standard functions can only return the base type results, and game developers are responsible for casting the results to the specific types defined in their games.

In Unreal, Cast is a safe conversion function used for casting data types. Here is the syntax:

```
ToType* Cast<ToType>(FromType* theObjectPointerOfFromType)
```

Let's break the syntax down:

- `ToType` represents the target type that you want to get after the conversion
- `FromType` represents the original type that you want to convert from
- The angular brackets (< and >) are C++ template expressions
- The function returns `nullptr` if the casting operation failed

In C++, a function can be declared to be generic with a template expression, which means that the function can accept and process different data types.

Recall the C++ `MYCPP_05` project in *Chapter 4*—the `Calculator` and `CalculatorEx` classes have two versions of the `Add` and `Subtract` functions:

```
int Add(int a, int b);
float Add(float a, float b);
int Subtract(int a, int b);
float Subtract(float a, float b);
```

To gain a clearer comprehension of programming with C++ template functions, please download the `MYCPP_06` project from the book's repository. In this sample project, the technique employed by the new `Calculator` and `CalculatorEx` classes involves merging and transforming the aforementioned `Add` and `Subtract` functions into two template functions, as follows:

```
template <typename T>
T Add<T>(T a, T b)
{
    return a + b;
}
template<typename T>
T Subtract<T>(T a, T b)
{
    Return a - b;
}
```

> **Note**
>
> Template function implementations should be moved into the header file so that the compiler knows to generate different versions of executions.

To use these two functions for additions and subtractions, you simply call them with the data type, as follows:

```
float f = calculator.Add<float>(1.5f, 2.0f); //f is 3.5f
float i = calculator.Subtract<int>(3, 2);    //i is 1
```

You should have a basic idea of why the Cast function is defined as a template function. Use the following code to cast the objects into Pangaea objects:

```
APangaeaPlayerController* pangaeaPlayerController =
        Cast<APangaeaPlayerController>(playerController);
APangaeaGameMode* pangaeaGameMode =
        Cast<APangaeaGameMode*> gameMode;
APangaeaGameStateBase* pangaeaGameState =
        Cast<APangaeaGameStateBase*>(gameState);
UPangaeaGameInstance* panGaeaGameInstance =
        Cast<UPangaeaGameInstance>(gameInstance);
```

The provided code snippet exemplifies the utilization of the Cast template function to convert base class pointers into the corresponding child class pointers within the game. It is essential to emphasize that casting operations are a commonly employed technique in **object-oriented programming** (OOP).

Summary

Throughout this chapter, you learned about Unreal Engine's basic gameplay framework classes, including AActor, APawn, ACharacter, APlayerController, AGameModeBase, AGameStateBase, and UGameInstance. By extending these classes, you can create the most needed elements for developing new games. Besides creating game actors, two important macros, UPROPERTY and UFUNCTION, were also introduced so that you can make actor properties and functions recognizable and work together with the engine. Your new classes' data can also be edited in the engine editor.

In this chapter, the following tasks were completed to enhance the game development process in Pangaea. New classes, including DefenseTower, Projectile, and PlayerAvatar, were created by inheriting from the Actor and Character classes. These classes were enriched with additional properties and functions to provide customization and unique behaviors.

Different methods for rebuilding *uprojects* after code changes were learned, ensuring that modifications were properly integrated into the game. Blueprint classes were then generated based on the C++ classes.

One important tool introduced as the final topic was the Cast template function, which facilitated the conversion of base class pointers into the appropriate child class pointers.

The upcoming chapter will guide you through the process of configuring the player avatar, including skeletal mesh setup, animations, and additional features. By the end, you will have the opportunity to replace the default player character with the newly customized player character.

6

Creating Game Actors

Game actors are the main elements of video games; they are controlled by either players (player actors or player characters) or AI controllers (non-player characters). A game actor may be represented by either a skeletal mesh (a warrior character, for instance) or a static mesh (a spaceship, for instance). The interactions between various game actors make up the gameplay.

This chapter will mainly guide you in creating your own player character (`PlayerAvatar`) for *Pangaea*. You will start by creating the animation instance class in C++, and then based on it, you will create the animation blueprint.

For the new animation blueprint, you will add the State Machine and define the states' animations. You will also set up the character's visual display in the character blueprint, define user inputs, and write code to implement the player controller and control the character states via the parameters defined in the animation instance class.

Once the player character setup is done, we will use it to replace the default player pawn for the **Game Mode**. After that, we will also create the defense tower and projectile actors.

The Unreal C++ scripting skills covered in this chapter are as follows:

- Setting up the player avatar
- Setting up the character's `SkeletalMeshComponent`
- Creating the player avatar's animation blueprint

Technical requirements

The code for this chapter can be found at `https://github.com/PacktPublishing/Unreal-Engine-5-Game-Development-with-C-Scripting/tree/main/Chapter06`.

Setting up the player avatar

Different actors are created for different purposes, so actors need different combinations of components. For example, a player character may need a `Camera` component attached to it for the top-down view, whereas a building needs a collision box to prevent pawns from moving through it.

Writing a script to add components to a new actor needs four steps:

1. Define a `private` variable that will hold the component pointer.
2. Add the `public` getter function, so that the component pointer can be retrieved outside of the class.
3. Include the added component's header file.
4. Instantiate the component in the class's constructor function.

To set up the top-down view for the game, we want to attach two more components (`SprintArmComponent` and `CameraComponent`) to the player avatar.

Adding SpringArmComponent and CameraComponent to PlayerAvatar

The `PlayerAvatar` class already has the components inherited from its parent class, `Character`, including a `Capsule` component, a `SkeletalMesh` component, a `CharacterMove` component, and an `Arrow` component.

To make it a player character for the top-down game, we need to add a `SpringArm` component and a `Camera` component to the `PlayerAvatar` class. The sprint arm should be bound to the origin of the character, and the camera is attached to the other end of the spring arm. You can see this illustrated here:

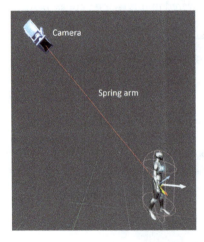

Figure 6.1 – Setting up the spring arm and camera for the player

In this case, we want to define the two variables, _springArmComponent and _cameraComponent, for storing the pointers of SpringArmComponent and CameraComponent. These two variables are marked with the UPROPERTY macro with the VisibleAnyWhere specifier so that they are visible in the editor.

To do this, add the following code to the end of APlayerAvatar in the header file:

```
private:

UPROPERTY(VisibleAnywhere, BlueprintReadOnly,
Category = "Camera",
meta = (AllowPrivateAccess = "true"))
class USpringArmComponent* _springArmComponent;

UPROPERTY(VisibleAnywhere, BlueprintReadOnly,
Category = "Camera",
meta = (AllowPrivateAccess = "true"))
class UCameraComponent* _cameraComponent;
```

Furthermore, apart from incorporating the properties into the class, our next objective is to include two getter functions, GetStringArmComponent and GetCameraComponent, in the public section of the class within the header file. For better performance, the functions can also be tagged with the FORCEINLINE macro:

```
FORCEINLINE USpringArmComponent* GetSringArmComponent() const
{
    return _springArmComponent;
}

FORCEINLINE UCameraComponent* GetCameraComponent() const
{
    return _cameraComponent;
}
```

> **About C++ inline functions**
>
> In C++ programming, an inline function's code is directly inserted at the call site instead of invoking a separate function call. This mechanism helps to improve the code execution performance by reducing the overhead of function calls. However, inline functions should only contain small and straightforward code snippets without complex flow control.

Now is the time to include `SpringArmComponent.h` and `CameraComponent.h` to avoid compilation errors. Please make sure to include `PlayerAvatar.generate.h` in the last `include` statement, so that Unreal's **Build** tool can properly parse Unreal macros:

```
#include "CoreMinimal.h"
#include "GameFramework/Character.h"
#include "GameFramework/SpringArmComponent.h"
#include "Camera/CameraComponent.h"
#include "PlayerAvatar.generated.h"
```

Next, we call the Unreal `UObject` class's member function, `CreateDefaultSubobject`, to instantiate components. Here is the function's syntax:

```
template<class TReturnType>
TReturnType * CreateDefaultSubobject
(
    FName SubobjectName,
    bool bTransient
)
```

Now, let's examine the function's parameters and its return value:

- `SubobjectName`: The name of the newly created component
- `bTransient`: The default value is `false`, which means the new component doesn't inherit the parent class defaults
- The `CreateDefaultSubobject()` function returns the pointer to the created object

Now, let's switch to `PlayerAvatar.cpp`, and add the instantiation code to the `PlayerAvatar` constructor function:

```
//Create the camera spring arm
_springArmComponent =CreateDefaultSubobject<USpringArmComponent>(
  TEXT("SpringArm"));
_springArmComponent->SetupAttachment(RootComponent);
_springArmComponent->SetUsingAbsoluteRotation(true);
_springArmComponent->TargetArmLength = 800.f;
_springArmComponent->SetRelativeRotation(
FRotator(-60.f, 0.f, 0.f));
_springArmComponent->bDoCollisionTest = false;

//Create the camera
_cameraComponent =CreateDefaultSubobject
  <UCameraComponent>(TEXT("Camera"));
_cameraComponent->SetupAttachment(_springArmComponent,
```

```
    USpringArmComponent::SocketName);
_cameraComponent->bUsePawnControlRotation = false;
```

The previous code, which includes some UObject member functions and variables, the TEXT macro, and the FRotator struct, is used to set up the camera view.

Let's take a closer look at the UObject member functions and variables first:

Function	Description
SetupAttachment (class: SceneComponent)	Attaches SceneComponent as a child to a parent scene component. This function can also attach the scene component to a designated skeletal mesh socket.
SetUsingAbsoluteRotation (class: SceneComponent)	Sets the flag to determine whether the scene component rotates with its parent or keeps its own absolute world rotation.
SetRelativeRotation (class: SceneComponent)	Sets the scene component's relative rotation to its parent.
Variable	**Description**
TargetArmLength (class: SprintArmComponent)	This variable determines the length of the spring arm component.
bDoCollisionTest (class: SprintArmComponent)	This flag indicates whether a collision test is applied to this spring arm component.
bUsePawnControlRotation (class: CameraComponent)	This flag indicates whether the rotation of the CameraComponent is controlled by the view/control rotation of the owning pawn.

Figure 6.2 – Component functions for setting up the player camera

In the previous code snippet, you may have observed the utilization of the TEXT macro to mark string literals as localization keys. The TEXT macro also guarantees appropriate data types for characters or string literals on different platforms. Different platforms may encode characters or strings according to different encoding standards, such as UTF8, UTF16, or UTF32, so using the TEXT macro can ensure characters and strings are encoded correctly and avoid type conversion errors. The value output from the TEXT macro can be assigned to FName (text strings optimized for higher performance), FText (text strings for displaying information to players that may be localized), and FString (regular mutable text strings) type variables.

The FRotator structure represents a 3D rotation. Like a class, a structure type is also an encapsulation of a set of variables and functions. The main difference is that a structure is a value type (a block of memory that stores some values), whereas a class is a reference type (stores an address pointing to a block of memory that stores some values). FRotator can be constructed with the Pitch, Yaw, and Roll parameters.

Besides creating components, we need to do some initializations, such as enabling ticking, setting controller constraints, and configuring the movement parameters, while constructing the player avatar.

Initializing the player avatar

For the player avatar, we still need to do some initializations in the class constructor to restrict the character's rotation and configure the MovementComponent.

Why do we do the initialization works inside the class's constructor rather than the BeginPlay function? This is because BeginPlay is only a game runtime function, whereas the added components and settings are needed during editing. So, runtime gameplay initializations can be placed in the BeginPlay function.

Let's add some code to APlayerAvatar::APlayerAvatar(). The first thing is to enable the ticking function, which indicates that the character's Tick function will be called with the game's frame update:

```
PrimaryActorTick.bCanEverTick = true;
```

Next, since the top-down gameplay requires the character to orient its running direction, the character's rotation should not be controlled by the player:

```
bUseControllerRotationPitch = false;
bUseControllerRotationYaw = false;
bUseControllerRotationRoll = false;
```

The last thing we want to set up is the character's movement component. The movement component should control the character to rotate toward its moving direction, and the rotation speed is 640 degrees around the y axis. In the meantime, the character needs to be constrained and snapped to the ground. Enter this code to set up this component:

```
auto characterMovement = GetCharacterMovement();
characterMovement->bOrientRotationToMovement = true;
characterMovement->RotationRate = FRotator(0.f, 640.f, 0.f);
characterMovement->bConstrainToPlane = true;
characterMovement->bSnapToPlaneAtStart = true;
```

Here, we used a new C++ keyword, `auto`, which allows the `CharacterMovement` variable's data type to be automatically determined by the `GetCharacterMovement` function's return value type.

Now the `PlayerAvatar` class is basically done. We can use it as a base class to create a new character blueprint so that we can set up and replace the game's default player pawn.

Setting up the character's SkeletalMeshComponent

To set up the player avatar, we set the skeletal mesh and its animations for the `BP_PlayerAvatar` blueprint and replace the default player pawn in the game mode settings.

But before the setting up works, we need to import the fancy character model and its animations.

Importing the character model

For the player avatar, we want to use another skeletal model to substitute the default skeletal mesh provided by Unreal Engine. You may choose to use any character models that are compatible with UE5. In this book, we are going to use the **Hero** model (the `.FBX` and `.TGA` files), which you can download from the book's GitHub repository under the `PangaeaAssets/Hero` folder, to set up the player avatar's **SkeletalMesh** component.

The next thing is to create a subfolder for importing the character assets. Here, we choose to create the new `Hero` folder under **All | Content | Characters**:

Figure 6.3 – Creating a new Hero folder

Now, perform the following steps to import the character assets:

1. The first step is to import the character model and its skeleton. Right-click on the **Hero** folder and choose **Import to | Game | Characters | Hero** in the pop-up menu. Find Hero.fbx in the **Import** dialog box and click the **Open** button to open the **FBX Import Options** window.

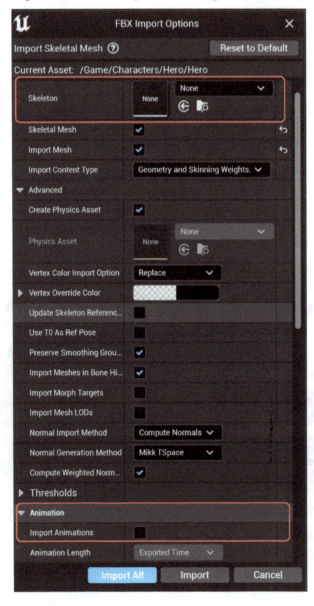

Figure 6.4 – Importing the Hero.fbx model and the skeleton

In the **FBX Import Options** window, do the following:

- The **Skeleton** field should be set to **None** – this means that, currently, the system doesn't have the skeleton for this model and the skeleton will be imported

- Make sure that the **Import Animations** option is unchecked, as the model file doesn't contain any animation information

- Then, click the **Import All** or **Import** button to start the process

Once the importing is done, close any popups and you should see four new imported assets: the skeletal mesh, the physics asset, the skeleton, and the material (**matHero**).

Figure 6.5 – The four imported assets from Hero.fbx

2. Now, carry out the same tasks as in *step 1* to import the two image files: T_CharA_COL.png and T_Char_NOR.png. These two images will be used as the color map and the normal map for the model's material:

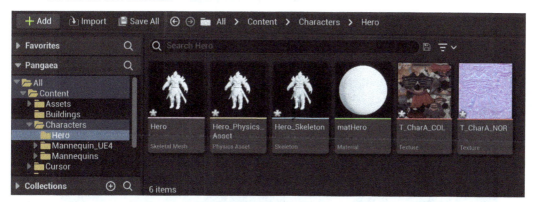

Figure 6.6 – The two imported color and normal maps

3. You may notice that all the model asset previews are in gray. We need to make some slight changes to the material so that the color and normal maps can be used as the skin to wrap the model:

    i.   Double-click **matHero** to open the Material Editor.

    ii. Add two **Texture Sample** nodes to the graph, and then connect the two nodes' **RGB** outputs to the material's **Base Color** and **Normal** input pins.

    iii. Select **T_CharA_COL** for the **Base Color** texture sample node.

    iv. Select **T_CharA_NOR** for the **Normal** texture sample node:

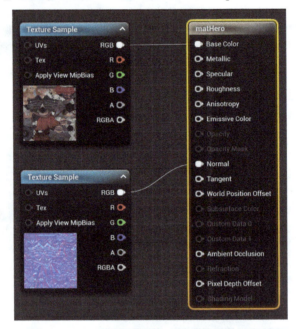

Figure 6.7 – Creating material for the Hero model

After clicking **Apply** and **Save** on the new material, the **Hero** assets should look colorful now:

Figure 6.8 – The Hero assets preview after changing the material

4. The downloaded `Hero.zip` file also comes with six `.fbx` files, which start with the prefix `Hero_Anim_`. You can import these animation files the same way as you did in *step 1*. This time, though, in the **FBX Import Options** window, the **Skeleton** field should be filled with **Hero_Skeleton** and the **Import Animations** option should be checked.

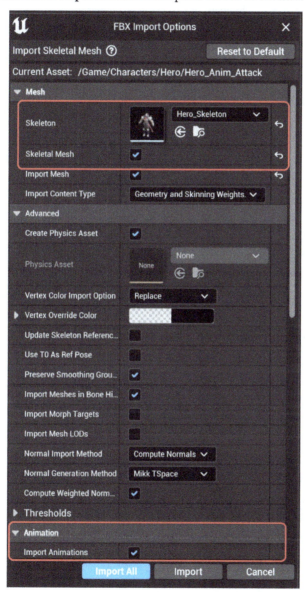

Figure 6.9 – Importing Hero animations

We've got the **Hero** assets all set now. Let's use them in the player avatar blueprint.

Using the Hero skeletal mesh in BP_PlayerAvatar

Open the BP_PlayerAvatar blueprint in the editor and click the **Viewport** tab; select the **Mesh** component from the components hierarchy, and choose the **Hero** skeletal mesh in the **Details** view. Then adjust the model's location by lowering it to **-90.0** units on the **Z** axis and rotating the model **-90.0** degrees around the **Z** axis as well. Now the character's feet are aligned with the ground and the character is facing the front:

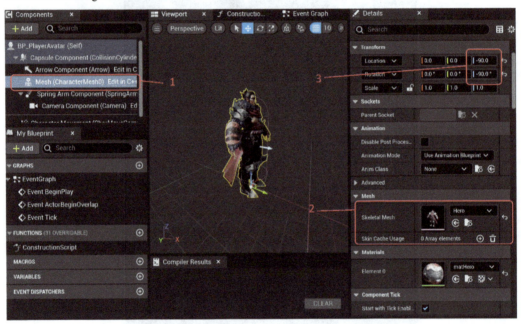

Figure 6.10 – Using the Hero skeletal model in BP_PlayerAvatar

Don't forget to save and compile the blueprint. Now, the player avatar blueprint is ready to be used to replace the default player pawn.

Replacing the game's player pawn

The game's default player pawn is set in the project's game mode in the **Project Settings** window. So, go to the editor's main menu and choose **Edit | Project Settings**. From here, you can then open the **Project Settings** window.

From the **Project** list group, choose **Map & Modes**. Then, from the drop-down list of the **Default GameMode** field, select **PangaeaGameMode**.

When **PangaeaGameMode** is picked as **Default GameMode**, expand **Selected GameMode**; you should find that the **Default Pawn Class** field is initially set as **BP_TopDownCharacter** and is grayed out, meaning it is not changeable.

The reason why the configuration fields under the **Selected GameMode** section are not changeable is that PangaeaGameMode is a C++ class, so the fields have to be assigned with appropriate values in the constructor of the class. To do that, open PangaeaGameMode.cpp and type in the following code:

```
static ConstructorHelpers::FClassFinder<APawn>
PlayerPawnBPClass(TEXT("/Game/TopDown/Blueprints/BP_
TopDownCharacter"));
if (PlayerPawnBPClass.Class != nullptr)
{
  DefaultPawnClass = PlayerPawnBPClass.Class;
}
```

The preceding code snippet tries to find the BP_TopDownCharacter asset with the engine's static function, ConstructorHelpers::FClassFinder, and assigns the result to the DefaultPawnClass system variable.

Since the path for finding the asset here is hardcoded, we simply change the asset name from BP_TopDownCharacter to BP_PlayerAvatar. Now, **Default Pawn Class** should be **BP_PlayerAvatar**, like so:

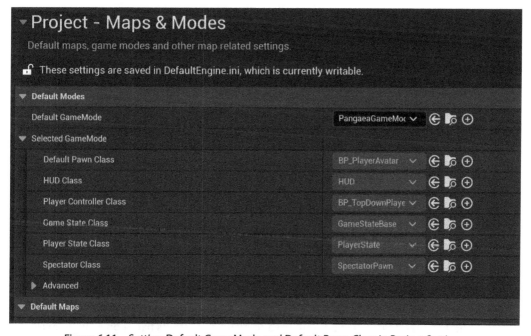

Figure 6.11 – Setting Default GameMode and Default Pawn Class in Project Settings

> **Attention**
>
> When exiting the running C++ program, do not **Stop** in Visual Studio. Instead, make sure that you first close your Unreal Editor window to ensure that all your imported assets are saved properly. Failure to do so might result in the loss of the currently imported assets and the settings.

Now, when you launch the game, the player character is the **Hero** character instead of the old top-down character:

Figure 6.12 – The Hero is now the player character

The character currently doesn't play animations when it moves around, so the next thing we want to do is to create the animation blueprint for the **Hero** character.

Creating the player avatar's animation blueprint

Once the character is set, the subsequent objective is to control the character animations. To accomplish this, we first create a C++ animation instance class, `PlayerAvataAnimInstance`, and then use it as the base class to create the animation blueprint, `ABP_PlayerAvatar`.

To keep things simple, we will assign the player character only the following animation states:

- **Locomotion**: This state has a **Blend Space 1D** that blends the *Idle*, *Walk*, and *Run* animations. A `Speed` float variable will be used to control the blending work.

> **Note**
>
> Unreal Engine's **Blend Space 1D** is a tool used to blend animations based on a single-axis parameter. It allows smooth transitions between animations by interpolating values along the specified axis. Animation poses, such as walk and run, are mapped along the speed axis.

- **Attack**: This state plays the *Attack* animation. A Boolean variable, `IsAttacking`, will be used to transit to and from this state.

- **Hit**: When the character gets hit, the player's avatar plays the *Hit* react animation.

- **Die**: If the character's HP equals or is lower than 0, the player's avatar plays the *Die* animation.

At the end of playing the *Attack*, *Hit*, and *Die* animations, the animations should fire an event to notify the animation instance class to set the flags to `false`, so that the **State Machine** transitions from its current state to the next appropriate state.

> **Note**
>
> A State Machine in Unreal is a graphical expression tool used in animation blueprints. It breaks the skeletal mesh into a series of states and controls the transitions among the animation states, based on user-defined rules.

As `PlayerAvatarAnimInstance` serves as a base class, our first focus is to create this C++ class.

Creating the PlayerAvatarAnimInstance class

To create the animation instance class, you should find the `AnimInstance` parent class in the **Add C++ Class** window. So, choose the **All Classes** tab button and type `Anim` in the search box; the `AnimInstance` class should appear in the list:

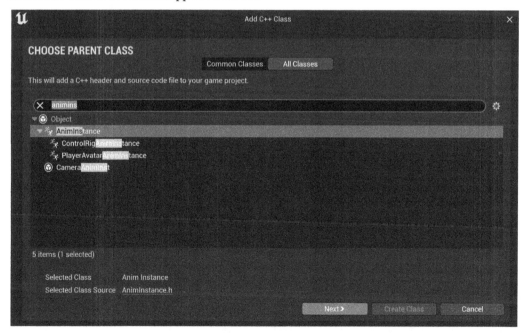

Figure 6.13 – Find the AnimInstance parent class for creating ABP_PlayerAvatar

Click **Next**, then type `PlayerAvatarAnimInstance` into the class's **Name** box. This example simply places the new source code files into the .../`Source`/`Pangaea` folder:

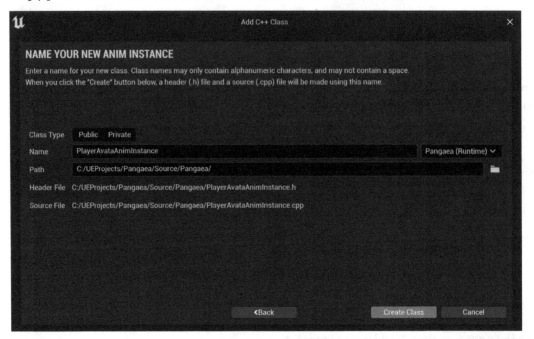

Figure 6.14 – Creating the PlayerAvatarAnimInstance class

Now, press the **Create Class** button, then reload and rebuild the project.

Next, open the `PlayerAvatarAnimInstance.h` file in VS – we need to write some code to add an `EPlayerState` enumeration type and two public variables: `Speed` and `State`. The `Speed` variable is used to control the interpolation of **Idle**, **Walk**, and **Run** animations when the current state is `Locomotion`. The enumeration type `State` variable can only be set with one of the enumerated values (`Locomotion`, `Attack`, `Hit`, and `Die`) to control and transit from one state to another.

> **Note**
>
> In C++ programming, an enumeration is a user-defined data type. The value of an enumeration can only be one of its defined range of values. The enumeration's values are explicitly expressed by constant names. To define an enumeration data type in C++, you should use the keyword `enum class` or `enum`. When scripting for Unreal, you should use the former.

The following code defines an `EPlayerState` enumeration type:

```
UENUM(BlueprintType)
enum class EPlayerState : uint8
{
    Locomotion,
    Attack,
    Hit,
    Die
};
```

Let's break this code down:

- This code defines a new enumeration data type called `EPlayerState` (we use the E prefix for the type name to indicate that this is an enumeration data type).

- The `EPlayerState` variables can only be assigned one of the four values: `Locomotion`, `Attack`, `Hit`, and `Die`.

- The `UENUM` macro has a `BlueprintType` specifier, which indicates that the `enum` data type is friendly to blueprints.

- We used `: uint8` to tell the compiler to reserve only 8 bits (1 byte) to store the value. Otherwise, the compiler will use 32 bits (4 bytes).

Now, we can define the `Speed` and `State` variables:

```
public :

UPROPERTY(EditAnywhere, BlueprintReadWrite)
float Speed;

UPROPERTY(EditAnywhere, BlueprintReadWrite)
EPlayerState State;
```

We also need to add a function, `OnStateAnimationEnds`, which takes care of changing the `State` when the *Attack*, *Hit*, or *Die* animation ends:

```
UFUNCTION(BlueprintCallable)
void OnStateAnimationEnds ();
```

Even though `State` is visible to blueprints and can be directly set in blueprints, we want to use this example to demonstrate how to use `BlueprintCallable` functions. In some cases (complex logic, for instance), writing C++ code is more effective and clearer than drawing spaghetti blueprint diagrams.

Here is the full code for the `PlayerAvatarAnimInstance.h` file:

```cpp
#include "CoreMinimal.h"
#include "Animation/AnimInstance.h"
#include "PlayerAvatarAnimInstance.generated.h"

UENUM(BlueprintType)
enum class EPlayerState : uint8
{
  Locomotion = 0,
  Attack,
  Hit,
  Die
};

UCLASS()
class PANGAEA_API UPlayerAvatarAnimInstance : public UAnimInstance
{
GENERATED_BODY()

public :

  UPROPERTY(EditAnywhere, BlueprintReadWrite)
  float Speed;

  UPROPERTY(EditAnywhere, BlueprintReadWrite)
  bool IsAttacking;

UPROPERTY(EditAnywhere, BlueprintReadWrite)
EPlayerState State;

UFUNCTION(BlueprintCallable)
void OnStateAnimationEnds ();

};
```

In the code, we declare the `OnStateAnimationEnds` function in the header file.

Then, the implementation of this function should be positioned in the `PlayerAvatarAnimInstance.cpp` source file:

```cpp
#include "PlayerAvatarAnimInstance.h"
#include "PlayerAvatar.h"
```

```
void UPlayerAvatarAnimInstance::OnStateAnimationEnds()
{
    if (State == EPlayerState::Attack)
    {
        State = EPlayerState::Locomotion;
    }
    else
    {
        auto ownerActor = this->GetOwningActor();
        auto playerAvatar =
        Cast<APlayerAvatar>(ownerActor);
    if (playerAvatar == nullptr)
    {
        Return;
    }
    if (State == EPlayerState::Hit)
    {
    if (playerAvatar->GetHealthPoints() > 0.0f)
    {
        State = EPlayerState::Locomotion;
    }
    else
    {
        State = EPlayerState::Die;
    }
    }
    else if (State == EPlayerState::Die)
    {
        //...
    }
    }
}
```

Looking into `PlayerAvatarAnimationInstance.cpp`, the first line of code, `#include "PlayerAvatar.h"`, provides the definition of `PlayerAvatar` and enables casting the animation instance's owner character from an `Actor` pointer to be a `PlayerAvatar` pointer, and as a result, the `PlayerAvatar::GetHealthPoints()` member function is accessible and can be invoked.

Figure 6.15 is a reference showing the function event graph, which does the equivalent process as the `OnStateAnimationEnds` function:

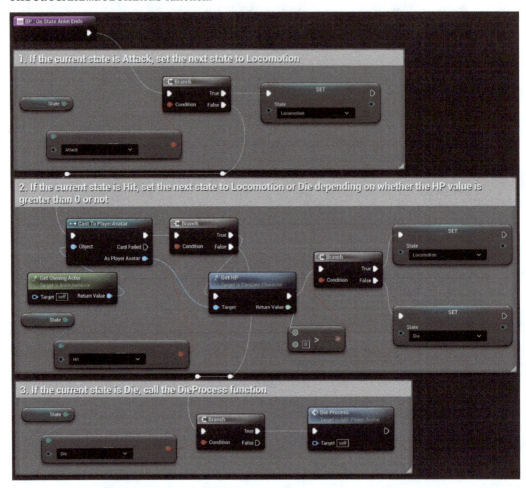

Figure 6.15 – The equivalent OnStateAnimationEnds function event graph

Once the `PlayerAvatarAnimationInstance` class has been implemented, we can proceed to create the `ABP_PlayerAvatar` animation blueprint. Based on this animation blueprint, we will define a State Machine that controls the character's animations by responding to state changes.

Creating the ABP_PlayerAvatar blueprint

To create the blueprint, right-click on the folder where you want to place the new blueprint (we chose the folder from **All** | **Content** | **TopDown** | **Blueprints**). Then select **Animation** | **Animation Blueprint** from the pop-up menu:

Figure 6.16 – Creating a new animation blueprint

The next thing to do is to choose which skeleton and what parent class are used to create this animation blueprint. Here, we chose to use the imported character skeleton and `PlayerAvatarAnimInstance` as the parent class:

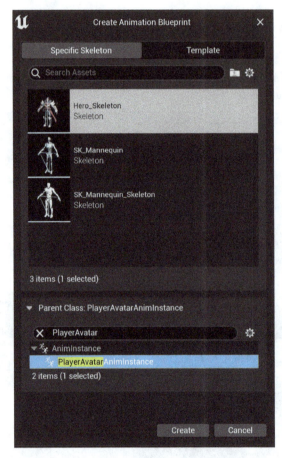

Figure 6.17 – Selecting the skeleton and parent class to create ABP_PlayerAvatar.png

After hitting the **Create** button, you should have the new animation blueprint in the `Blueprints` folder. Rename it `ABP_PlayerAvatar`.

The next step is to create a State Machine on the `ABP_PlayerAvatar` animation blueprint.

Creating the State Machine on ABP_PlayerAvatar

The Unreal State Machine allows developers to set up character states and the corresponding state animations. By using a State Machine, we can easily use the `Speed` and `State` variables to control the transitions between states.

Since this book's focus is on C++ scripting, we will not go into the detailed steps of how to create a State Machine. If you don't have knowledge of the State Machine in Unreal, please visit `https://docs.unrealengine.com/5.0/en-US/state-machines-in-unreal-engine/` to learn the fundamentals and search online for Unreal State Machine video tutorials.

Here in this book, we just list the results for all the steps for creating a **Blend Space 1D** asset, adding a new State Machine to the animation Blueprint, and adding states and state transitions. So, let's get started:

1. Create a **Blend Space 1D** asset, `HeroBlendSpace1D`, for the `Locomotion` state. Then add the idle, walk, and run animations to the blending space line, and change **Horizontal Axis Name** to `Speed`.

Figure 6.18 – Creating the BlendSpace1D asset

2. Add a new State Machine to the animation blueprint and connect the State Machine's output to **Output Pose**.

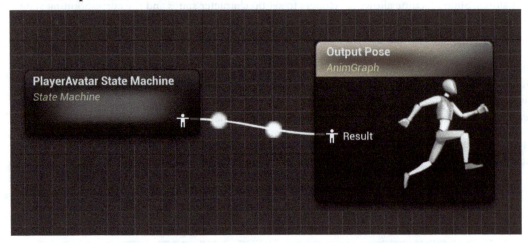

Figure 6.19 – Adding a State Machine to ABP_PlayerAvatar

3. For the State Machine, add four new states and name them Locomotion, Attack, Hit, and Die. Then, connect them as shown in the following figure:

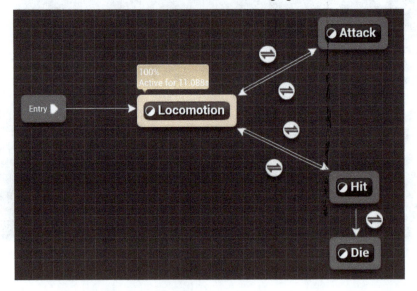

Figure 6.20 – The ABP_PlayerAvatar State Machine

4. Double-click on the **Locomotion** node to enter the state's graph editor, add the **HeroBlendSpace1D** and **Speed** nodes, and connect the nodes as shown in *Figure 6.21*. The **Speed** node can be found in the **All Possible Actions** list; this getter node returns the value of the `PlayerAvatarAniminstance::Speed` variable.

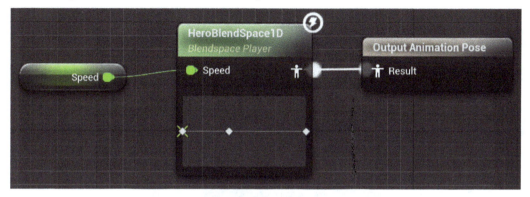

Figure 6.21 – Adding HeroBlendSpace1D to the Locomotion state

In the state graph, the **HeroBlendSpace1D** node utilizes the **Speed** input value to interpolate the idle, walk, and run animations for the **Output Animation Pose** node.

5. Set up the `Attack`, `Hit`, and `Die` states by simply adding appropriate animations to the states, and then connect the animation outputs to the **Output Animation Pose** nodes.

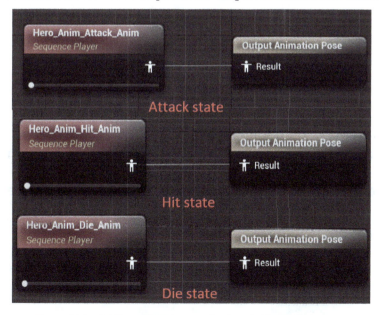

Figure 6.22 – Adding animations to the Attack, Hit, and Die states

6. For each state transition (arrowed connection) created in *Figure 6.20*, set the state transition conditions by adding the **State** and the **State** comparison nodes. Each comparison between `PlayerAvataAnimInstance::State` and the specific state triggers the transition when the comparison result is true.

Figure 6.23 – Adding conditions to transition between states

7. Compile and save the animation blueprint.

8. Now, we can associate it with our player avatar. Open `BP_PlayerAvatar`, select the **Mesh** component, and set **Use Animation Blueprint** for **Animation Mode** and **ABP_PlayerAvatar** for **Anim Class**.

The last thing we want to do is to sync the character's movement speed with the `Speed` variable of `PlayerAvatarAnimInstance` for the `Locomotion` state to blend the animations.

Syncing the movement speed with the animation instance

To sync the movement speed, we can write some code in `PlayerAvatar.cpp`. For every tick, the code will do the following:

-   First, get the animation instance from the character's skeletal mesh and cast it to be a `UPlayerAvatarAnimInstance` class pointer; then the result is assigned to the `playerAvatarAnimInst` variable.

-   Second, read the `Velocity` vector from the character's movement component.

- Finally, calculate the `Velocity` vector's length and assign it to the `Speed` variable of `playerAvatarAnimInst`.

Here is the new code for the `Tick()` function:

```
void APlayerAvatar::Tick(float DeltaTime)
{
Super::Tick(DeltaTime);

  UPlayerAvatarAnimInstance* animInst =
      Cast<UPlayerAvatarAnimInstance>(
GetMesh()->GetAnimInstance());
animInst->Speed =
      GetCharacterMovement()->Velocity.Size2D();
}
```

The `Tick()` function's `DeltaTime` parameter indicates the elapsed time in seconds since the previous frame tick. For example, if a game ticks at 100 frames per second, the `DeltaTime` value is 0.01f.

The `velocity` value we get here is an `FVector` struct data type, which indicates the velocities along with the *x*, *y*, and *z* axes in the world coordinate system. But what we are interested in is the combined speed of *x* and *y*. So, what we need is the return value of `Fvector::Size2D()`.

The `FVector` structure type is an expression of the 3D space mathematical term **vector**. In Unreal, an `FVector` value is composed of three float-type components (`X`, `Y`, and `Z`). `FVector` is used for 3D space representations (location, Euler angle rotation, etc.) and vector calculations (movement, cosine of the angle between two vectors, the normal vector of two vectors, etc.). For your reference, we list some frequently used `FVector` functions here:

Function Name	Is Static?	Description
`CrossProduct`	Yes	Calculates two vectors' cross product, which is the normal vector of the surface formed by the original two vectors
`DotProduct`	Yes	Calculates two vectors' dot product; the result equals a multiplication of the lengths of the two vectors and the cosine value of the two vectors' angles
`Distance`	Yes	Calculates the distance between the two locations
`DistSquare`	Yes	Calculates the squared distance between the two vector locations
`RotateAngleAxis`	No	Rotates this vector along an axis for degree angles and returns the rotated vector
`Rotation`	No	Returns the rotation (`FRotator`) of this vector
`Size`	No	Returns the 3D (X, Y, Z) length of this vector

Size2D	No	Returns the 2D (X, Y) length of this vector
FVector Operators	No	+, -, ==, ! =, and so on.

Figure 6.24 – List of some useful FVector functions

Visit the Unreal Engine documentation site for more details on vectors: `https://docs.unrealengine.com/4.26/en-US/API/Runtime/Core/Math/FVector/`.

Upon completing this chapter, you should have your own player character, the **Hero**, for the *Pangaea* game. You can now launch the game and make your **Hero** walk around:

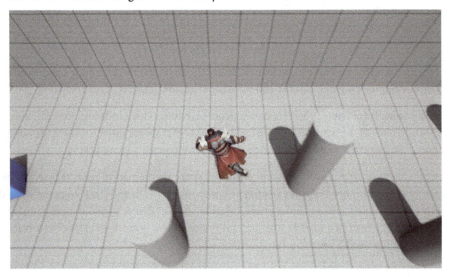

Figure 6.25 – Pangaea's new player character

Summary

Having completing this chapter, you should now be capable of creating your own game-player character from scratch.

First, you learned how to set up the top-down camera view by adding a Camera component and a SprintArm component to the PlayerAvatar class. The camera is attached to the spring arm, so it is placed at the top-down view position and moves with the character. In the PlayerAvatar constructor, you not only created the SpringArm and Camera components but also initialized the settings for the character as well as its CharacterMovement component.

Once BP_PlayerAvatar was set up, it was used to substitute the default top-down game character.

You then created the PlayerAvatarAnimInstance class and defined the two variables (Speed and State). Based on that, you also created the animation blueprint and defined the State Machine for leveraging the character's state animations.

Finally, you wrote the code for the APlayerAvatar::Tick() function to sync the character's velocity with the Speed variable of the animation instance.

In the next chapter, we will show you how to set up user inputs and control the player character. We will also create an enemy character and the AI controller that controls the enemy's behaviors.

7
Controlling Characters

In *Chapter 6*, we created our player character, the `PlayerAvatar` class, and the `BP_PlayerAvatar` blueprint. The next thing we want to do is to control the player character. Since the engine automatically generated the default `PangaeaPlayerController` class when the project was created, moving the player character is already functional. However, that is not enough because we want more control over the player character.

In this chapter, you will learn how to add a new input action to the action map, handle an input event to trigger an attack, as well as how to add a notify to the timelines of attack, hit, and die animations.

This chapter will also introduce the concept of garbage collection, which is the engine's important mechanism for memory management; the code for destroying characters demonstrates how to manually trigger garbage collection.

Additionally, you will learn how to create an enemy character and control the enemy with `AIController`. This entails enabling the enemy to sense and chase the player and make decisions to attack, as well as handling destruction upon being defeated.

Both `EnemyController` and `PangaeaPlayerController` move the controlled character with the pathfinding algorithm, based on the **NavMesh** system, so we will explain the essential concepts of pathfinding and demonstrate the practical setup process within the game map.

This chapter covers the following topics:

- Controlling the player character to attack
- Destroying actors
- Creating the enemy character
- Testing the game

Technical requirements

The code for this chapter can be found at https://github.com/PacktPublishing/Unreal-Engine-5-Game-Development-with-C-Scripting/tree/main/Chapter07.

Controlling the player character to attack

When you created the *Pangaea* project, the engine automatically generated the PangaeaPlayerController class so that you had basic control over the player character. To understand how PangaeaPlayerController controls the player, let's open the PangaeaPlayerController.cpp file in **Visual Studio (VS)** and look at the code.

First of all, the SetupInputComponent() function binds the OnSetDestinationPressed() and the OnSetDestinationReleased() event handler functions to the **SetDestination** action defined in the project's input settings. These two event handler functions call the movement functions, StopMovement and SimpleMoveToLocation, to move the character toward the next new destination.

Can we add a new attack action to the system and hook it up to our own handler function? The answer is yes. Let's start by defining the new attack action.

Adding the Attack action to the action map

In Unreal, PlayerController is defined as an interface that interprets player input, enabling the controlled pawn to respond accordingly to the player's commands and actions.

PangaeaPlayercontroller is a child class of the engine's PlayerController class, so it inherits all the useful variables and functions from the parent. For example, you can periodically call the following three *checking* functions of PlayerControllers to directly check whether an input key (e.g., the *Esc* key) was just pressed down, was just released up, or is held down, like so:

- bool WasInputKeyJustPressed(Ekeys::Escape)
- bool WasInputKeyJustReleased(Ekeys::Escape)
- bool IsInputKeyDown(Ekeys::Escape)

It is easy and simple to capture user inputs directly from specific input keys, but directly capturing key inputs is not very flexible to support multiple input devices. Suppose that a game needs to support firing bullets by pressing either the spacebar, the left mouse button, or a controller trigger; in such cases, you will need to call the previous functions three times to check the states of all three devices.

Unreal provides another solution to handle user inputs, known as **input mapping**. The idea is that you can define **Action** or **Axis** mapping groups in your game project, and then bind any number of input methods to the action or axis. For our *Pangaea* game, we can add a new action, **Attack**, and bind the right mouse button to it.

To accomplish this, begin by navigating to **Edit | Project Settings** from the main menu in Unreal Editor. Next, locate **Input** under the **Engine** group, and press the + button to add a new **Attack** action to the **Action Mappings**. Finally, add **Right Mouse Button** and **Space Bar** as the inputs for this action:

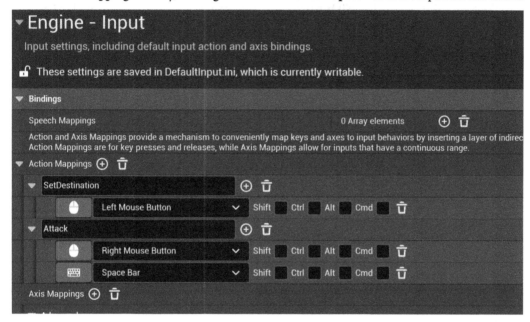

Figure 7.1 – Adding the new Attack Action Mapping

Once you've added the new Action Mapping, you can close the **Project Settings** window and try binding the process function to the Attack action in your code.

Binding the handler function to the Attack action

The checking functions you have just learned about need to check the input device states upon every tick, which is not a very efficient process. The **Input** mapping mechanism, on the other hand, handles player input by events. Only when a player input is captured will it trigger the event, and then the event handler functions are called. In that case, we can write a function and bind it to the Attack action.

Open PangaeaPlayerController.h and add the following code to declare the new member function:

```
void OnAttackPressed();
```
Then, in PangaeaPlayerController.cpp, find the SetupInputComponent function, and add the following code at the end to bind OnAttackPressed() function to the Attack action:
```
    InputComponent->BindAction("Attack",
IE_Pressed,
```

```
this,
&APangaeaPlayerController::OnAttackPressed);
```

The next question is, what will happen when the Attack event is triggered? The answer is very straightforward. We want to check whether PlayerAvatar can attack; if it returns true, PlayerAvatar's Attack() function is called.

Implementing the OnAttackPressed() action handler function

Open PangaeaPlayerController.cpp and add the following code block to implement the OnAttackPressed() function:

```
void APangaeaPlayerController::OnAttackPressed()
{
    auto playerAvatar = Cast<APlayerAvatar>(GetPawn());
    if (playerAvatar->CanAttack())
    {
        playerAvatar->Attack();
    }
}
```

Let's break down this code:

- The GetPawn function returns the pointer of the pawn currently controlled

- The Cast function casts the returned APawn* pointer to be an APlayerAvatar* pointer

Next, we need to implement the two CanAttack and Attack functions. The former checks whether the player has cooled down and is allowed to start another attack, whereas the latter simply restarts the countdown, which will then cause the player to start playing the **Attack** animation.

Implementing the CanAttack() and Attack() functions

As you may recall, we previously declared the CanAttack and Attack functions during the creation of the PlayerAvatar class. Now is the time to add the code to get them to work. Let's open PlayerAvatar.cpp and enter the following code block for the implementation of the CanAttack function:

```
bool APlayerAvatar::CanAttack()
{
UPlayerAvatarAnimInstance* animInst = cast<UPlayerAvatarAnimInstance>(
    GetMesh()->GetAnimInstance());
return (_AttackCountingDown <= 0.0f &&
animInst->State == EPlayerState::Locomotion);
}
```

Let's break down this code:

- The `GetMesh` function returns the pointer of the character's **SkeletalMesh** component
- The `GetAnimInstance` function returns the pointer of the animation instance associated with the skeletal mesh
- The `cast` function casts the returned `UAnimInstance*` pointer to be the `UPlayerAvatarAnimInstance*` pointer
- This `CanAttack()` function returns `true` if the `Attack` state counting down time is over and its current state is `Locomotion`; otherwise, it returns `false`

Now, we will implement the `Attack()` function:

```
void APlayerAvatar::Attack()
{
    _AttackCountingDown = AttackInterval;
}
```

The `Attack` function simply resets the `Attack` state cooling down timer. Since the character's `Tick` function checks and changes the character's state to `Attack` when the values of `_AttackCountingDown` and `AttackInterval` are equal, setting `_AttackCountingDown` with `AttackInterval` can start playing the **Attack** animation.

There are two more tasks we want to do in the character's `Tick` function:

- Change the `PlayerAvatar` animation state to `Attack` so that the state machine starts playing the **Attack** animation
- Reduce `_AttackCountingDown` so that the `CanAttack` function can return `true` after `_AttackCountingDown` reaches 0

These two tasks are implemented like so:

```
void APlayerAvatar::Tick(float DeltaTime)
{
    Super::Tick(DeltaTime);
    auto animInst = Cast<UPlayerAvatarAnimInstance>(
        GetMesh()->GetAnimInstance());
        animInst->Speed =
GetCharacterMovement()->Velocity.Size2D();

  if (_AttackCountingDown == AttackInterval)
  {
    animInst->State = EPlayerState::Attack;
  }
```

```
if (_AttackCountingDown > 0.0f)
{
  _AttackCountingDown -= DeltaTime;
}
}
```

Now, compile and launch the game by clicking your mouse's right button – the character should attack. The problem now is that your character plays the **Attack** animation, never terminates, and transits back to the Locomotion state. Let's solve the problem when playing non-loop animations.

Processing non-loop animations

To solve the problem of playing and terminating non-loop animations (**Attack**, **Hit**, and **Die**), we need to add notifications on the timelines of the non-loop animations in the animation blueprint. Then, we can write and hook event handler functions to do the corresponding processes. Let's get started.

Adding animation end notifies

To add a *notify* on to the **Attack** animation, we will use the Hero_Anim_Attack animation in the Content Drawer/All/Content/Characters/Hero folder. Double-click on it to open an animation in the **Animation Sequence Editor**, and then perform the following steps to add a new notify on the animation timeline (see *Figure 7.2*):

1. Right-click on the **Notifies** track at a position near the endpoint of the animation.

2. Select **Add Notify…** from the pop-up menu.

3. Select **New Notify…** from the submenu.

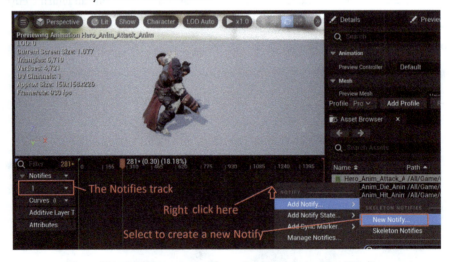

Figure 7.2 – Adding a new notify on the Attack animation timeline

4. Input `AttackEnds` as the name of the new notify. The notify tag should be shown on the animation timeline:

Figure 7.3 – The AttackEnds notify is added to the animation timeline

5. You can follow the same process to add notifies for the *hit* and *die* animations, using `HitEnds` and `DieEnds` as the notify names, respectively.

Once all the animation notifies are set up, we can handle the corresponding events and transit the character's animation state to the next appropriate state.

Handling the notify events on the animation blueprint

We just added notifies to trigger the animation events while playing the animations; now, the animation blueprint can handle those events and do the processes.

On the event graph of `ABP_PlayerAvatar`, follow these steps:

1. Right-click on the **Event Graph** area and search for `AttackEnds`, `HitEnds`, and `DieEnds` to add the events.

2. Right-click on the **Event Graph** area and search for the `OnStateAnimationEnds` function. This is defined in the `AplayerAvatarAnimInstance` class.

3. Connect all the events' outputs to the `OnStateAnimationEnds` function.

The Event Graph blueprint should look like the following figure:

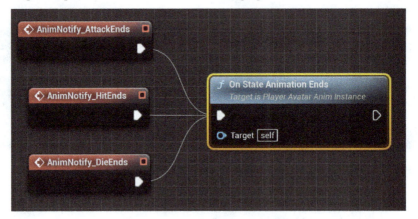

Figure 7.4 – The Event Graph to handle animation end events

We have added *notifies* for the *attack*, *hit*, and *die* animations and handled the animation notifies in the blueprint to call the `OnStateAnimationEnds` event handler function. To accomplish this mechanism, we have to implement the event function and get the system functional.

Implementing the OnStateAnimationEnds function

The `OnStateAnimationEnds` function's main job is to transit the animation state to the next state properly. The function can be called when the current animation state is either `Attack`, `Hit`, or `Die`, so they need to be processed based on the following logic:

- If the function is called when the current state is `Attack`, meaning that the **Attack** animation just ended, then it should go back to the `Locomotion` state.

- If the function is called when the current state is `Hit`, we need to check whether the `HealthPoints` value of the character is greater than 0, and then we turn the next state to `Locomotion`; otherwise, the character is killed and should transit to the `Die` state to start playing the `Hero_Anim_Die` animation.

- If the function is called when the current state is `Die`, this means that the character is dead. We will want to call the `APlayerAvatar::DieProcess()` function to remove and destroy the character from the scene.

To implement this, here is the code for the `OnStateAnimationEnds()` function:

```
void UPlayerAvatarAnimInstance::OnStateAnimationEnds()
{
    if (State == EPlayerState::Attack)
    {
```

```
    State = EPlayerState::Locomotion;
  }
  else
  {
    auto playerAvatar =
    Cast<APlayerAvatar>(this->GetOwningActor());
  if (State == EPlayerState::Hit)
  {
  if (playerAvatar->GetHealthPoints() > 0.0f)
  {
    State = EPlayerState::Locomotion;
  }
  else
  {
    State = EPlayerState::Die;
  }
  }
  else if (State == EPlayerState::Die)
  {
    playerAvatar->DieProcess();
  }
  }
}
```

We just implemented the `PangaeaPlayerController`, allowing players to control the hero's attacks. In the next section, we will address the issue of appropriately destroying actors upon their demise.

Destroying actors

Game actors may need to be destroyed when they are killed or no longer needed in a game map. The *Pangaea* player character, for example, is killed when `HealthPoints` reaches 0; once it reaches 0, then we want to remove the player character from the map and kick the player from the game.

To complete this task, we will do three things in the `DieProcess()` function:

- Stop the character from ticking. Setting the `PrimaryActorTick.bCanEverTick` variable as `false` does this.

- Call the `K2_DestroyActor()` function to destroy the character. This Unreal actor API destroys the actor and marks the occupied memory for garbage collection, resulting in the release of memory

- Call the engine's `ForceGarbageCollection` function to force the engine to perform garbage collection.

To implement this, we can insert the following three lines of code within the curly braces of the `DieProcess` function:

```
void APlayerAvatar::DieProcess()
{
    PrimaryActorTick.bCanEverTick = false;
    Destroy();
    GEngine->ForceGarbageCollection(true);
}
```

The code block was used to demonstrate how to explicitly stop an actor's ticking and request a garbage collection for the system.

However, in this case, you can opt for the following simplified line of code, as it achieves the same outcome, the difference being that we let the engine handle the processes internally:

```
void APlayerAvatar::DieProcess()
{
    Destroy();
}
```

When an actor is instantiated, the engine allocates a block of memory to store the actor's information, and in the meantime, the returned actor pointer stores the starting address of the memory block. When actors are destroyed, their memory blocks are not immediately released. Only when a garbage collection starts will all the destroyed objects' memory be released for new memory allocations.

Garbage collection is a memory management mechanism that Unreal uses to recycle unused memory blocks. Unreal internally manages memory allocations and deallocations. Garbage collection can be started under three conditions:

- Garbage collection starts periodically
- When a system is running out of memory
- When a developer manually forces garbage collection to start

In the case of destroying the player character in our game, we force the engine to recycle the released memory because the game is over, and we want the engine to clear the unused memory before exiting.

Garbage collection is a very expensive process, so you should try to avoid triggering it during gameplay; it may impact the frame rate and cause noticeable chops.

We are basically done with the player character setups. The next thing we want to do is create the enemy character. This will make the game fun to play.

Creating the enemy character

Creating the enemy character is similar to creating the player character. First, we create the `AEnemy` class, which inherits from `ACharacter`. Second, we create the `BP_Enemy` blueprint, with `AEnemy` as the parent class. Third, we create the `ABP_Enemy` animation blueprint, which is identical to `ABP_PlayerAvatar`.

The main difference between the enemy and player characters is the controller. The `EnemyController` class inherits from the engine's `AIController` and will make decisions to move to the target and attack.

The enemy will share the hero's model and animations, so to distinguish between the two, enemy models will use a gray material, which puts all the enemies in gray (see *Figure 7.5*):

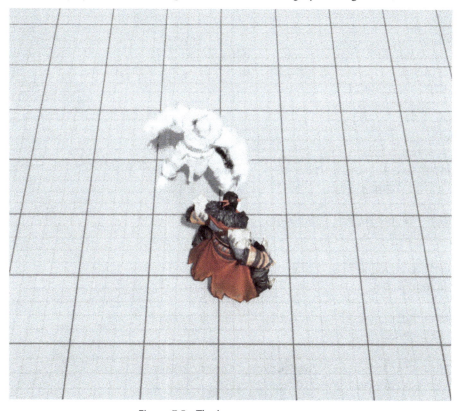

Figure 7.5 – The hero versus an enemy

Before we start, we want to clarify that we will apply some iterative processes in the development of the game. While creating the enemy character, we will copy some code and borrow some assets from the player character. The benefit of doing this is that it makes it easier and clearer for you. It is great if you notice some redundant code and feel it would be better to combine such instances of code – this means you have a good understanding of the term **code refactoring**, which we shall cover in *Chapter 9*.

Creating the Enemy class

In the same way that we created the `APlayerAvatar` class, we can create a new `AEnemy` class, which also inherits from `ACharacter`. The newly generated `Enemy.h` and `Enemy.cpp` files should still be placed under the path of `C++ Classes/Pangaea`.

Type the following code into the `Enemy.h` header file:

```cpp
#pragma once

#include "CoreMinimal.h"
#include "GameFramework/Character.h"
#include "Enemy.generated.h"

UCLASS()
class PANGAEA_API AEnemy : public ACharacter
{
    GENERATED_BODY()

public:
AEnemy();

UPROPERTY(EditAnywhere, Category = "Enemy Params")
int HealthPoints = 100;

UPROPERTY(EditAnywhere, Category = "Enemy Params")
float Strength = 5.0;

UPROPERTY(EditAnywhere, Category = "Enemy Params")
float Armor = 1;

UPROPERTY(EditAnywhere, Category = "Enemy Params")
float AttackRange = 200.0f;

UPROPERTY(EditAnywhere, Category = "Enemy Params")
float AttackInterval = 3.0f;

protected:
virtual void BeginPlay() override;

int _HealthPoints;
float _AttackCountingDown;
APawn* _chasedTarget = nullptr;
```

```
public:
virtual void Tick(float DeltaTime) override;

UFUNCTION(BlueprintCallable,
Category = "Pangaea|Enemy",
meta = (DisplayName = "Get HP"))
int GetHealthPoints();

UFUNCTION(BlueprintCallable, Category = "Pangaea|Enemy")
bool IsKilled();

UFUNCTION(BlueprintCallable, Category = "Pangaea|Enemy")
bool CanAttack();

UFUNCTION(BlueprintCallable, Category = "Pangaea|Enemy")
void Chase(APawn* targetPawn);

void Attack();
void Hit(int damage);
void DieProcess();
bool IsAttacking();

private:
UPROPERTY(VisibleAnywhere,
BlueprintReadOnly,
meta = (AllowPrivateAccess = "true"))
class UPawnSensingComponent* PawnSensingComponent;
};
```

Note that the `AEnemy` class header file code looks quite similar to that of the `APlayerAvatar` class, except for the following differences:

- The initial values of the `HealthPoints`, `Strength`, `Armor`, and `AttackInterval` variables are set differently so that the enemy is weaker and slower than the player character. These values can also be fine-tuned in the character's details window in the editor.

- We added the `_chasedTarget` variable to cache the pointer to the current target, which, in this case, is the player character.

- We added `PawnSensingComponent`. This component works like a radar that checks whether the player character is within the sensing scope. When the player character is detected, the `_chasedTarget` variable is set to the player character. If the player character is out of range, this variable is set as `nullptr`.

- We added a new Chase(APawn* targetPawn) function. This function is marked as a BlueprintCallable function. When PawnSensingComponent triggers the OnSeePawn event, the blueprint calls the Chase function to set the target, and then the enemy character starts to chase the player character.

- Since enemy characters don't accept player input, we don't want to override the SetupPlayerInputComponent function as we did for the PlayerAvatar class.

Now, here is the code for Enemy.cpp:

```cpp
#include "Enemy.h"
#include "Perception/PawnSensingComponent.h"
#include "GameFramework/CharacterMovementComponent.h"
#include "EnemyController.h"
#include "EnemyAnimInstance.h"

AEnemy::AEnemy()
{
PrimaryActorTick.bCanEverTick = true;

PawnSensingComponent =
CreateDefaultSubobject<UPawnSensingComponent>(
TEXT("PawnSensor"));
}

void AEnemy::BeginPlay()
{
Super::BeginPlay();
_HealthPoints = HealthPoints;
}

void AEnemy::Tick(float DeltaTime)
{
Super::Tick(DeltaTime);

auto animInst = Cast<UEnemyAnimInstance>(
GetMesh()->GetAnimInstance());
animInst->Speed =
GetCharacterMovement()->Velocity.Size2D();

if (_AttackCountingDown == AttackInterval)
{
animInst->State = EEnemyState::Attack;
}
```

```cpp
if (_AttackCountingDown > 0.0f)
{
_AttackCountingDown -= DeltaTime;
}


if (_chasedTarget != nullptr &&
animInst->State == EEnemyState::Locomotion)
{
auto enemyController =
Cast<AEnemyController>(GetController());
enemyController->MakeAttackDecision(_chasedTarget);
}
}


int AEnemy::GetHealthPoints()
{
return _HealthPoints;
}


bool AEnemy::IsKilled()
{
return (_HealthPoints <= 0.0f);
}


bool AEnemy::CanAttack()
{
auto animInst = GetMesh()->GetAnimInstance();
auto enemyAnimInst = Cast<UEnemyAnimInstance>(animInst);
return (_AttackCountingDown <= 0.0f &&
enemyAnimInst->State == EEnemyState::Locomotion);
}


void AEnemy::Chase(APawn* targetPawn)
{
auto animInst = GetMesh()->GetAnimInstance();
auto enemyAnimInst =
Cast<UPlayerAvatarAnimInstance>(animInst);
if (targetPawn != nullptr &&
enemyAnimInst->State == EEnemyState::Locomotion)
{
auto enemyController =
```

```cpp
Cast<AEnemyController>(GetController());
enemyController->MoveToActor(targetPawn, 90.0f);
}
_chasedTarget = targetPawn;
}

void AEnemy::Attack()
{
GetController()->StopMovement();
_AttackCountingDown = AttackInterval;
}

void AEnemy::Hit(int damage)
{
_HealthPoints -= damage;

auto animInst = GetMesh()->GetAnimInstance();
auto enemyAnimInst =
Cast<UEnemyAnimInstance>(animInst);
enemyAnimInst->State = EPlayerState::Hit;

if (IsKilled())
{
DieProcess();
}
}

void AEnemy::DieProcess()
{
PrimaryActorTick.bCanEverTick = false;
K2_DestroyActor();
GEngine->ForceGarbageCollection(true);
}
```

In comparison with the PlayerAvatar class, the two additional things that the Enemy class does is create PawnSensingComponent and implement the Chase function.

In the class constructor, the CreateDefaultSubobject function is called to create the component.

The Chase function has one parameter, which passes in the target pawn to be chased. The function first gets the enemy character's controller, which is a subclass of AIController, and then it calls the MoveToActor function to move the enemy to the target player character.

You may have noticed that **VS** at this point complains about some errors, but don't panic – we haven't created the `EnenmyController` class and implemented the `Chase` function yet. Obviously, the enemy needs its controller, which controls the enemy to chase and attack in the game, and the animation instance – let's create them.

Creating the EnemyController class

In Unreal, you have a few different options to control **non-player characters** (**NPCs**). For instance, the **Behavior Tree** (**BT**) and the **Blackboard** can be used to design NPCs' behaviors. However, here, we will create a subclass of `AIController` and control the enemy character by writing scripts.

Even though the demo game only needs `EnemyController` to implement one function, which makes the decision to attack, it shows you how the new controller works, and you can always add more complex decision-making logic with advanced AI algorithms later.

Now, let's create the new C++ class, `EnemyController`, which inherits from the `AIController` class. The newly generated `EnemyController.h` and `EnemyController.cpp` files should be placed under `C++ Classes` | `Pangaea`.

The source code for `EnemyController.h` should be like this:

```
#pragma once

#include "CoreMinimal.h"
#include "AIController.h"
#include "EnemyController.generated.h"

UCLASS()
class PANGAEA_API AEnemyController : public AAIController
{
GENERATED_BODY()
public:
void MakeAttackDecision(APawn *targetPawn);
};
```

And the source code for `EnemyController.cpp` should be like this:

```
#include "EnemyController.h"
#include "Enemy.h"

void AEnemyController::MakeAttackDecision(APawn* targetPawn)
{
auto controlledCharacter = Cast<AEnemy>(GetPawn());
auto dist = FVector::Dist2D(
targetPawn->GetActorLocation(),
```

```
GetPawn()->GetTargetLocation());

  if (dist <= controlledCharacter->AttackRange
    && controlledCharacter->CanAttack())
  {
    controlledCharacter->Attack();
  }
}
```

What the MakeAttackDecision function does is check whether the distance between the chased target and the enemy is shorter than the enemy's attack range. If the target is within the attack scope and the enemy can attack, it calls the Attack function of the controlled character, which is the owner enemy.

In addition to making attack decisions, movement is another essential function for enemies. To accomplish this task, we will utilize AIController, provided by the game engine.

The AIController class is a useful tool to control NPCs. In *Pangaea*, the enemies are NPCs, so their movements are controlled by the AIController instances. Enemies can call the MoveToActor or MoveToLocation function of AIController to navigate within game maps.

To enhance your understanding of the navigation and the underlying theory, we want to introduce a bit more about AIController's navigation functions as well as how the pathfinding works with the NavMesh:

- MoveToActor: Moves the possessed pawn to an actor's location.

- MoveToLocation: Moves the possessed pawn to a destination location. Movement information is passed through multiple parameters.

- MoveTo: Moves the possessed pawn to a destination location. Use the FAIMoveRequest struct as the parameter to pass the movement information.

- StopMovement: Stops and aborts the current movement.

NavMesh

If you have no experience of NavMesh generation setups, you can search online for the official documentation and video tutorials, or refer to the official documentation on the Unreal navigation system web page: https://docs.unrealengine.com/5.0/en-US/navigation-system-in-unreal-engine/.

The `AIController`-controlled movement is based on an algorithm called **heuristic search** or the **A* pathfinding algorithm**. What the pathfinding algorithm does is find the shortest path from a start location to a destination location on the map. The map should have navigation information, which includes the nodes and the connections between nodes. Unreal builds the navigation information from the map's meshes, and the generated navigation data is stored as the maps' NavMesh.

When any `Move` function of the `AIController` class is called, it uses the owner pawn's current location as the starting point and the given destination location to generate the shortest walkable path, based on the map's NavMesh. The controlled pawn then moves along with the path toward its destination.

To ensure that the found path is valid and game pawns navigate properly, the map's NavMesh should be set up accordingly. To generate NavMesh, the basic requirement entails dragging and dropping `NavMeshBoundsVolume` into your game level and making it encompass the designated area for NavMesh generation (see *Figure 7.6*).

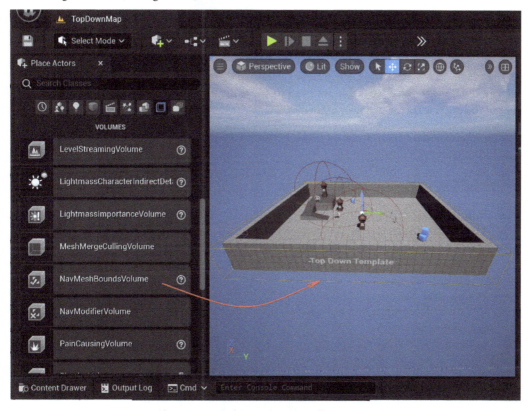

Figure 7.6 – Placing NavMeshBoundsVolume for NavMesh generation

To toggle showing or hiding the NavMesh, you can press *P* on your keyboard. The green shape covering the ground areas shows the NavMesh:

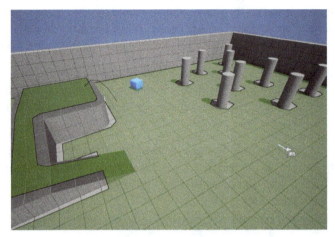

Figure 7.7 – The NavMesh is shown on the map after pressing the P key

Now, the enemies have an `EnemyController`, allowing them to sense, chase, and attack the player hero. The subsequent objective is that we need to create an animation instance and an animation blueprint to synchronize the behavior animations.

Create the `UEnemyAnimInstance` class by inheriting from the `AnimInstance` class. The header file should contain the following code:

```
#pragma once

#include "CoreMinimal.h"
#include "Animation/AnimInstance.h"
#include "EnemyAnimInstance.generated.h"

UENUM(BlueprintType)
enum class EEnemyState : uint8
{
    Locomotion,
    Attack,
    Hit,
    Die
};

UCLASS()
class PANGAEA_API UEnemyAnimInstance : public UAnimInstance
{
```

```
        GENERATED_BODY()


public:

UPROPERTY(EditAnywhere,
BlueprintReadWrite, Category = "Enemy Params")
float Speed;

UPROPERTY(EditAnywhere,
BlueprintReadWrite, Category = "Enemy Params")
EEnemyState State;

UFUNCTION(BlueprintCallable)
void OnStateAnimationEnds();


};
```

The provided code is similar to that in `PlayerAvatarAnimInstance.h`, which should already be familiar to you. Now, let's proceed by entering the following code into `EnemyAnimInstance.cpp` for the function implementations:

```
#include "EnemyAnimInstance.h"
#include "Enemy.h"

void UEnemyAnimInstance::OnStateAnimationEnds()
{
  if (State == EEnemyState::Attack)
  {
    State = EEnemyState::Locomotion;
  }
  else
  {
    auto enemy = Cast<AEnemy>(GetOwningActor());
  if (State == EEnemyState::Hit)
  {
  if (enemy->GetHealthPoints() > 0.0f)
  {
    State = EEnemyState::Locomotion;
  }
  else
  {
    State = EEnemyState::Die;
  }
```

```
    }
    else if (State == EEnemyState::Die)
    {
      enemy->DieProcess();
    }
    }
  }
```

In this case, the code for the `EnemyAnimInstance` class looks quite similar to that for the `PlayerAvatarAnimInstance` class, except for the redefined `EEnemyState` and casting the owning actor as the Enemy class pointer. Instead of applying class generalization, there are two reasons to retain the redundant code for the `EnemyAnimInstance` class:

- Maintaining consistency avoids confusion, as it is easier to understand the code
- Making the class extendable is useful for future developments

Now, we can create the animation blueprint, `ABP_Enemy`.

Creating the ABP_Enemy animation blueprint

For the *Pangaea* game, the enemy animation blueprint can be created in the same way that we created the `ABP_PlayerAvatar` blueprint in *Chapter 6*, except that its parent class is the `EnemyAnimInstance` class.

Perform the following steps to create `ABP_Enemy`:

1. Create a new animation blueprint.
2. Select `EnemyAnimInstance` as the parent class.
3. Select **Hero Skeleton** so that the enemy can share the animations of the hero.
4. Create a state machine on `AnimGraph`.
5. Add the states of `Locomotion`, `Attack`, `Hit`, and `Die` to the state machine.
6. For each state, add the corresponding animation.
7. Add state transitions and set the transit conditions.
8. Hook up the events of `AttackEnds`, `HitEnds`, and `DieEnds` to the `OnStateAnimationEnds` handler function.

You can see the result in *Figure 7.8*:

Figure 7.8 – The State Machine for ABP_Enemy

After completing the preceding tasks, we have everything ready to create the enemy blueprint.

Creating the BP_Enemy blueprint

We will download and import the enemy assets before the creation of the blueprint. Please follow these steps:

1. Download the assets from the Git repository under the `Enemy` folder, and then import the model into the project.

2. Create an `Enemy` folder under `All|Content|Characters`.

3. Import the `Enemy.FBX` model into the new `Enemy` folder.

Figure 7.9 – Importing the Enemy model

Now, we can create the animation blueprint.

4. Under the `All|Content|Topdown|Blueprints` folder, right-click and choose the `Blueprints|Blueprint` class to create `BP_Enemy`.

Then, we need to do some configuration work.

5. Set up the enemy's **SkeletalMesh** component. Choose **Enemy** for the skeletal mesh and **ABP_Enemy** for the **Anim Class** fields:

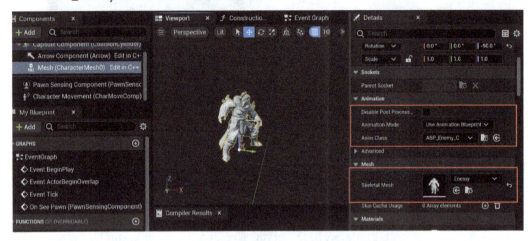

Figure 7.10 – Setting up the enemy's SkeletalMesh component

6. Set the **Pawn Sensing Component** values for visual detection, such as **Hearing Threshold**, **Sight Radius, Sensing Interval, Enable Sensing Updates, See Pawns**, and **Hear Noises**.

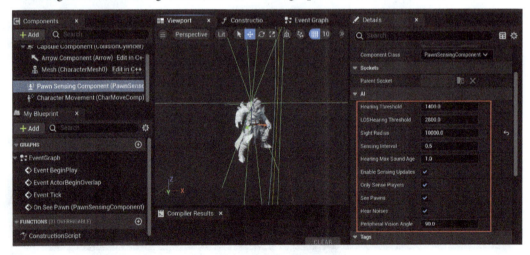

Figure 7.11 – Setting the Pawn Sensing Component values

7. Select the **Character Movement** component. Then, on the **Details** panel, adjust **Max Walk Speed** to 500.0 for the enemy to make it move a little bit slower than the hero, which means that it is possible for the player to run away from enemies.

Figure 7.12 – Setting Max Walk Speed to 500

8. In the BP_Enemy blueprint, add an **On See Pawn** event to trigger the Chase() function.

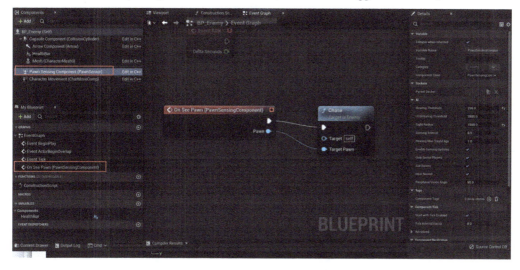

Figure 7.13 – Adding the On See Pawn event

Now ,we can compile and save the project, as it is the time now to test the game and see how things work together.

Testing the game

Drag BP_Enemy from the **Content Drawer** and place it in the game level. Then, launch the game. The enemy should chase the player character when the player character runs within its vision range:

Figure 7.14 – The enemy chases the hero

When the enemy runs close enough to the player character, it starts playing the **Attack** animation:

Figure 7.15 – The enemy attacks the hero

Summary

In this chapter, we showed you how to control both a player and an NPC character. First, we added the Attack action to the input action map and added the event handling function to the player controller. Then, we bound the function to the Attack action to control the hero to attack.

We learned how to add notifies to the animation timelines so that when the non-loop animations end, the animation blueprint can capture the notifications and properly transit to the next state, or if the character is killed, the DieProcess function is called to release memory and force garbage collection.

We imported new assets for the enemy character and created the Enemy class, the blueprint, and the animation blueprint, as we did for the player character. We also added PawnSensingComponent to the Enemy class so that enemies can detect whether the player character is within their vision range.

The main difference from the player character we made for the enemy was that we created the EnemyController class, which derives from the AIController class. Then, we used the controller to move the enemy based on Unreal's NavMesh. Also, we wrote code to make the Attack decision for enemies.

Now, our game has a hero and an enemy. They both can walk around and attack each other, but how do they really interact with each other? That should be handled by collisions. In the next chapter, we will explore adding collision components to actors and handling collision events.

8

Handling Collisions

Collision detection is a useful and effective mechanism to deal with interactions between game actors. The basic idea is that when one actor collides with another actor, a collision event is triggered to notify the two actors about the collision. The game developer can write a script to handle the events and perform the corresponding processes.

In this chapter, we will first introduce you to the basics of the collision detection system, the types of collision components, the collision events, and some of the **collision presets** that we will use for the development of *Pangaea*.

Throughout this learning process, you will acquire several skills. This includes adding colliders to actors and meshes to enable accurate collision detection, configuring collision presets, handling overlap events for weapon pickup, and casting rays to perform projectile hit checks.

To demonstrate the usage of the robust mathematical tool `FVector`, we will focus on simulating the trajectory and movement of projectiles. By utilizing `FVector` and employing vector calculations, you can enhance your understanding and proficiency in the aforementioned fundamental mathematical skills, empowering you to apply them throughout your own game development journey.

This chapter covers the following topics:

- Understanding collision detection
- Setting the collision detection presets
- Using collisions for game interactions

Technical requirements

The code for this chapter can be found at `https://github.com/PacktPublishing/Unreal-Engine-5-Game-Development-with-C-Scripting/tree/main/Chapter08`.

Understanding collision detection

Game collision detection is a system to detect when two or more game objects overlap or interact with each other in a game world. The process of collision detection is based on mathematical calculations that check whether two shapes intersect with each other (overlap detection) or whether a line goes through a surface (ray casting detection).

To detect actor collisions, Unreal offers certain types of simple collision components that use simple shapes to deal with collision detections:

- `UCapsuleComponent`
- `UBoxComponent`
- `USphereComponent`

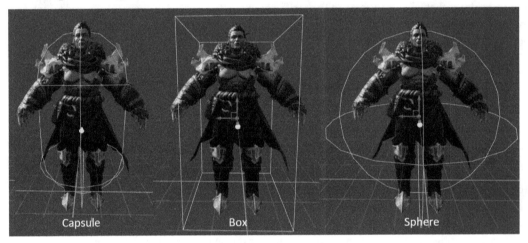

Figure 8.1 – Simple shape collision components

Besides using the collision components, another option to add simple collision shapes to static meshes is to add them via the **Mesh Editor**. Double-click on any mesh (model) on the **Content Drawer** to open the **Mesh Editor**. You should find the **Collision** menu item in the Blueprint Editor menu or on the toolbar:

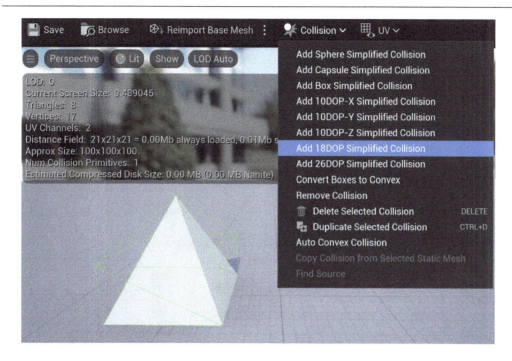

Figure 8.2 – Adding a simplified collision shape for the pyramid in the Mesh Editor

From here, choose an appropriate collision shape for the mesh from the drop-down list. *Figure 8.2* shows that the **18DOP Simplified Collision** (a bounding volume with 18 axis-aligned planes) item is selected to create the collision shape for the pyramid.

If no simple collision shape is added to a mesh, the mesh itself can be used for collision detections. The benefit is that you get very accurate collision detections, but the trade-off is a loss in performance.

The Unreal collision detection system provides interfaces for the gameplay program to sense what just happened in the game world. Messages of detected collisions are sent to the gameplay program in the form of collision events.

There are three basic collision events – OnBeginOverlap, OnEndOverLap, and OnHit:

- OnBeginOverlap events are triggered when an actor or a component enters a trigger area. For example, the player character enters a pickup area covered by the collision component and triggers the OnBeginOverlap event of the weapon, and then the weapon's event handler function attaches the weapon to the player character.

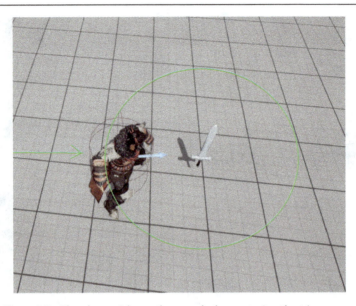

Figure 8.3 – The player picks up the sword when entering the trigger area

- OnEndOverlap events are triggered when an actor or a component leaves a trigger area. For example, when the player character walks close to a door, the door automatically opens, whereas when the player leaves the door area, the OnEndOverlap event is triggered, and the door closes.

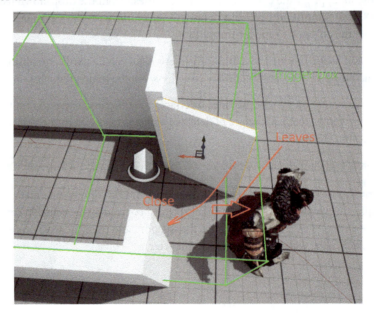

Figure 8.4 – Closing the door when the player leaves

- `OnHit` events occur when an actor moves and hits a solid object – for example, a wall. In this case, the actor is blocked from moving further so that overlapping with the wall is prevented.

Figure 8.5 – The player is stopped when colliding with the wall

When an actor has collidable components, such as `StaticMeshComponent`, `SkeletalMeshComponent`, `BoxComponent`, `SphereComponent`, and `CapsuleComponent`, the following actor collision events can be handled by event handler functions:

- `OnActorBeginOverlap`

- `OnActorEndOverlap`

- `OnActorHit`

A collidable component also has its own collision event interfaces that can be handled with event handler functions:

- `OnComponentBeginOverlap`

- `OnComponentEndOverlap`

- `OnComponentHit`

When designing a new actor, you should add appropriate collidable components to the actor and then hook up event handler functions to the actor or component collision events. For example, both a player character's skeletal mesh and an added capsule component are collidable components added to the character, and it's preferable to choose the capsule component instead of the **SkeletalMesh** component for better performance.

Collidable components can be configured independently for different use cases. They can be defined as a blocker, an overlap, a trigger, and so on. We can set the collision properties for a collidable component through its collision presets.

Setting the collision presets

Open an actor in the Actor Editor and find the **Collision Presets** section in the **Details** panel (see *Figure 8.6*). The collision presets property of a collidable component has only one drop-down box, which lists the optional collision presets. The settings of a selected preset can be expanded and viewed by clicking the **Play** button to the left of the **Collision Presets** label in the actor's **Details** tab panel.

The drop-down box has a number of options that you can choose from, but we will just introduce a few of them here (for more information, you can go to the official document site: `https://docs.unrealengine.com/5.0/en-US/collision-response-reference-in-unreal-engine/`):

- **Custom**: All the collision settings can be freely set by the developer.
- **NoCollision**: This collider has no collision.
- **BlockAll**: This collider blocks all actors in the scene. If a moving actor collides with this actor, the `OnActorHit` and `OnComponentHit` events are triggered.
- **OverlapAll**: This collider overlaps with all actors in the scene. If a moving actor enters or leaves this collider's scope, the actor and the component's `BeginOverlap` and `EndOverlap` events are triggered.
- **BlockAllDynamic**: This option is like **BlockAll** but only blocks pawns, cameras, and vehicles.
- **OverlapAllDynamic**: This option is like **OverlapAll** but only overlaps with pawns, cameras, and vehicles.

- **OverlapOnlyPawn**: This option is like **OverlapAll** but only overlaps with pawns.

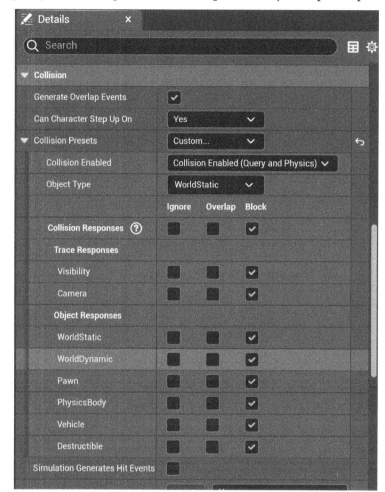

Figure 8.6 – The Collision Presets settings

If you select the **Custom** preset, you can change the collision setting according to your actual need. The custom preset is a matrix that has nine rows of object types and three checkbox columns titled **Ignore**, **Overlap**, and **Block**:

- **Ignore**: The physics body of this object type is ignored by this collision component.

- **Overlap**: The physics body of this object type is not blocked and generates `Overlap` events. This collision component acts like a trigger.

- **Block**: The physics body of this object type is blocked and generates `Hit` events. This collision component acts like a collider.

When two actors overlap with each other, the **Generate Overlap Events** checkboxes of these two actors' collision components must be checked to trigger the `Overlap` events.

Similarly, the `Hit` events are triggered only when the **Simulation Generates Hit Events** checkbox is checked.

The following table shows the collision settings of the actors in *Pangaea*:

Actor class	Collision component	Preset type	Generate overlap?	Simulation generates hit events?
PlayerAvatar	Capsule	Pawn	Yes	No
	SkeletalMesh	NoCollision	No	No
Enemy	Capsule	Pawn	Yes	No
	SkeletalMesh	NoCollision	No	No
Weapon (sword, axe, or hammer)	StaticMesh	OverlapAll	Yes	No
DefenseTower	SphereCollision	OverlapAllDynamic	Yes	No
	StaticMesh	BlockAllDynamics	Yes	No
Projectile (fireball)	StaticMesh	NoCollision	No	No

Figure 8.7 – Pangaea actor collision preset settings

The provided list of collision settings for different actor types facilitates the game's interaction mechanism within the game:

- Both the `PlayerAvatar` and `Enemy` actors have two collision components – `Capsule` and `SkeletalMesh`. Their `Capsule` components are set as `Pawn` so that they can be blocked by other `BlockAll` or `BlockAllDynamics` collision components.

- The `Capsule` component of the `PlayerAvatar` and `Enemy` actors' **Generate Overlap** option is set to `true` to enable overlapping events to trigger when overlapped by other `OverlapAll` and `OverlapAllDynamics` collision components.

- The `StaticMesh` component of the `Weapon` actors is set as an `OverlapAll` type of trigger. The **Generate Overlap** option is set to `true` so that when the weapon overlaps an actor, it can handle the `BeginActorOverlap` event to deal damage to the target.

- `DefenseTower` actors have two collision components – `SphereCollision` and `StaticMesh`. The `SphereCollision` component is set as `OverlapOnlyPawn` and the **Generate Overlaps** option is set to `true` so that when any pawn enters or leaves the sphere scope, the tower starts or stops firing at the invading target.

`Projectile` actors do not rely on collision components to detect hitting targets. Instead, we will utilize the engine's ray-tracing system. Therefore, the `StaticMesh` component is set as `NoCollision`. We have introduced the knowledge required for handling collisions; now, let's apply it to develop the *Pangaea* game interactions.

Based on the completed setups of actor collisions, we can proceed to develop the game interactions, which will ultimately deliver an enjoyable and engaging gameplay experience.

Using collisions for game interactions

To make it fun to play the game, we want to add the following gameplay features:

- Two types of weapons (sword and axe) can be placed in the game level. The player should walk through a weapon to pick it up. If the player character already has a weapon, when they pick up another weapon, the old weapon is dropped.

- When an enemy is activated, spawn a hammer weapon for the enemy.

- Defense towers should be placed in the game level. If the player character enters and stays within the attack range of the tower, the tower fires at the player character by spawning fireballs.

- A fireball moves along its firing direction and checks whether it hits the target. If the target is hit, it deals damage and destroys itself. Otherwise, the fireball flies for three seconds and destroys itself.

Before we start working on the game, we need to import the weapon and defense tower assets.

Downloading and creating the weapon, defense tower, and fireball actors

To get started, download the assets from the GitHub repository under the `/PangaeaAssets/Weapons` and `/PangaeaAssets/DefenseTower` folders. Here, you will find the following assets:

- **Axe**: `Axe.FBX` and `Axe_c.TGA`

- **Hammer**: `Hammer.FBX` and `Hammer_c.TGA`

- **Sword**: `Sword.FBX` and `Sword.TGA`

- **Defense tower**: `DefenseTower.FBX` and `DefenseTower_c.TGA`

Import these assets into the *Pangaea* project and create the `BP_Axe`, `BP_Hammer`, `BP_Sword`, and `BP_DefenseTower` blueprint. It is recommended to organize the files into suitable folders:

- Put the weapon assets, including the mesh, the texture, and the material files, into the `All/Content/Assets/Weapons` folder

- Put the defense tower assets, including the mesh, the texture, and the material files, into the `All/Content/ Assets/Buildings` folder

- Put all the new blueprints into the `All/Content/Assets/TopDown/Blueprints` folder

All the new materials simply have a single **Texture Sample** node, which takes in a diffuse texture map and connects to the **Base Color** pin of the **Result** node.

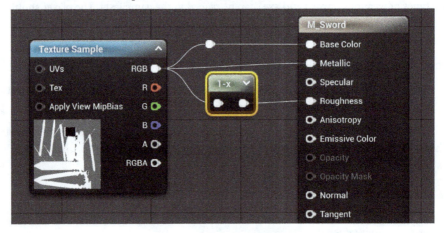

Figure 8.8 – The Sword material

This material simply connects the texture map to the **Metallic** and **Roughness** pins. Then, we add a `1 - x` node to revert the value to make the blade reflective, which you can see in *Figure 8.9*:

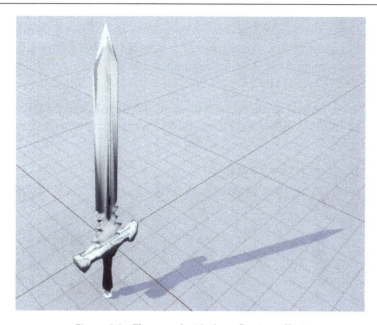

Figure 8.9 – The sword with the reflection effect

We also need to create another M_Fireball material for the fireball (see *Figure 8.10*); the fireball actor can use the Shape_Sphere mesh, which comes from **StarterContent**.

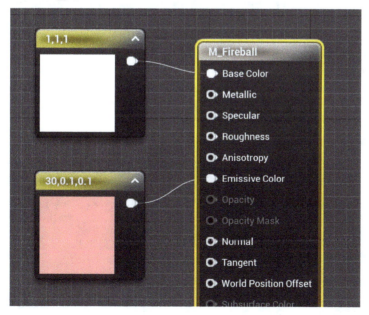

Figure 8.10 – Creating the M_Fireball material

To create the BP_Axe, BP_Hammer, and BP_Sword blueprints, we will inherit them from the Weapon class. These blueprints will utilize the **Axe**, **Hammer**, and **Sword** meshes for visual representation. Additionally, we need to assign the M_Axe, M_Hammer, and M_Sword materials to the respective weapon meshes.

Moreover, the **Collision Presets** settings should be set to **OverlapAll**, and the **Generate Overlap Events** box should be checked. This configuration enables the triggering of overlap events whenever weapons come into contact with any pawns or static meshes.

Figure 8.11 is an example of the creation of BP_Sword.

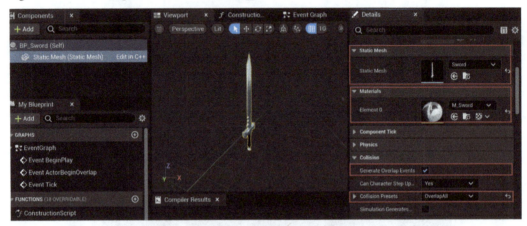

Figure 8.11 – Creating the BP_Sword blueprint

We then create the defense tower blueprints by inheriting from the DefenseTower class. Use the DefenseTower mesh for the visualization and the M_DefenseTower material. For the **Sphere Component**, set **Sphere Radius** to 800.0 and **Collision Presets** to **OverlapAllDynamic**, making sure that the **Generate Overlap Events** box is checked:

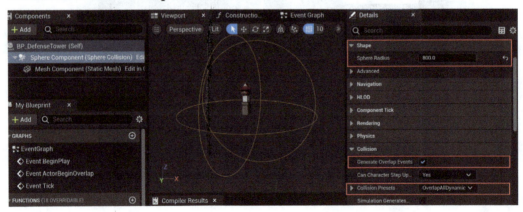

Figure 8.12 – BP_DefenseTower blueprint – Sphere Component setup

After setting up the **Sphere Component,** the **Mesh Component**'s collision presets should be set as **BlockAllDynamics**:

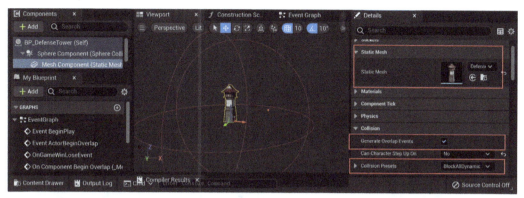

Figure 8.13 – BP_DefenseTower blueprint – Mesh Component setup

We create the fireball blueprint by inheriting from the `Projectile` class. The `Shape_Sphere` mesh and the `M_Fireball` material are used for the visualization. The mesh needs to be scaled down to `0.2` to fit the game, and **Collision Presets** should be set as **NoCollision**.

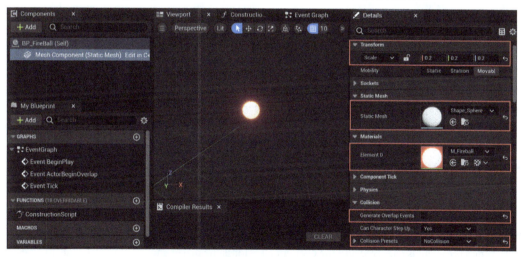

Figure 8.14 – Creating the BP_Fireball blueprint

Now, drag and drop some defense towers and weapons into the game level. For example, we can arrange two towers, one sword, and one axe on the ground :

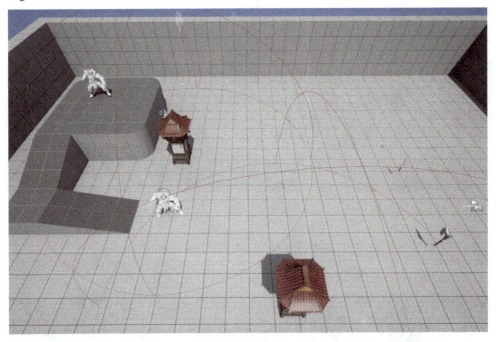

Figure 8.15 – Placing actors into the game level

Here, the player is spawned on the right side of the scene. A sword and an axe are placed in front of the player character to pick up. Two defense towers are on the path to the stage. One enemy is the guard of the ramp, and another enemy character represents the boss.

The next thing we want to do is to write some code to get things to work.

Picking up weapons

To make the actor pick up weapons, we let the `Weapon` class handle the `OnActorBeginOverlap` event. When the event is triggered and the overlapped actor is the player character, the weapon attaches itself to the character.

You may be asking, which part of the character will be the parent of the attached weapon? The solution is that we can add a socket to the right-hand bone of the character's skeleton, and then attach the picked weapon to that socket.

To add the socket, open `Hero_Skeleton` in the editor. Find the **hand_r** bone on the skeleton, right-click on it, and choose **Add Socket** from the pop-up menu to add a new socket under the bone. Rename the socket `hand_rSocket`. With the socket selected, change its **Relative Location** and **Relative Rotation** values according to those shown in the following figure:

Figure 8.16 – Adding the socket to attach a weapon

To add a **Preview Asset** to the socket, right-click on the socket and select **Add Preview Asset** from the pop-up menu. From there, you can choose a weapon of your choice as the **Preview Asset**.

> **Note**
>
> The **Preview Asset** will not be visible during gameplay. Its purpose is solely to assist in adjusting the relative location and rotation of the socket during the editing phase.

Once done setting the socket, we can add the declaration of the `OnWeaponBeginOverlap()` event handler function to the `Weapon` class in the header file:

```
UFUNCTION()
void OnWeaponBeginOverlap(AActor* OverlappedActor,
AActor* OtherActor);
```

To ensure proper functionality, we must bind the `OnWeaponBeginOverlap` function to the `OnActorBeginOverlap` delegate event of the weapon actor within the `BeginPlay()` function. This binding establishes the connection between the event and the corresponding function to handle weapon overlap events.

The delegate's macro function, AddDynamic, can then be used to bind event handler functions:

```
OnActorBeginOverlap.AddDynamic(this,
                    &AWeapon::OnWeaponBeginOverlap);
```

The first parameter, this, represents the weapon actor itself, and the second parameter, &AWeapon::OnWeaponBeginOverlap, represents the handler event function's address.

In the OnWeaponBeginOverlap function, we utilize the two functions from the AActor class, AttachToComponent and DetachFromActor. These functions are employed to handle the process of picking up and attaching the weapon to the overlapped player, as well as dropping the previously picked-up weapon.

Let's take a close look at the AttachToComponent function first:

```
AttachToComponent(Hero->GetMesh(),
FAttachmentTransformRules::SnapToTargetIncludingScale,
FName("hand_rSocket"));
```

Here is an explanation of the code:

- The first parameter passed is the player character's skeletal mesh
- The second parameter is an engine-defined enum value, which tells the function to snap the weapon to the target node on the mesh's skeleton
- The third parameter provides the target socket name

Next, let's examine the DetachFromActor function:

```
DetachFromActor(FDetachmentTransformRules::KeepWorldTransform);
```

The only parameter of this function is an engine-defined Enum value, which tells the function to drop the weapon at its current location.

Based on the previously explained pickup process and the related functions, here is the complete code for Weapon.h and Weapon.cpp, provided for your reference.

This is the weapon.h code:

```
#pragma once

#include "CoreMinimal.h"
#include "GameFramework/Actor.h"
#include "GameFramework/Character.h"
#include "Components/SphereComponent.h"
#include "Weapon.generated.h"
```

```
UCLASS()
class PANGAEA_API AWeapon : public AActor
{
GENERATED_BODY()
public:
AWeapon();

UPROPERTY(VisibleAnywhere, BlueprintReadWrite)
ACharacter* Holder = nullptr;

UPROPERTY(EditAnywhere, Category = "Weapon Params")
float Strength = 10;

protected:
virtual void BeginPlay() override;

UPROPERTY(VisibleAnywhere, BlueprintReadOnly)
UStaticMeshComponent* _StaticMesh;

UFUNCTION()
void OnWeaponBeginOverlap(AActor* OverlappedActor,
AActor* OtherActor);

bool IsWithinAttackRange(float AttackRange,
AActor* Target);

public:
virtual void Tick(float DeltaTime) override;
};
```

Here is an explanation of the code:

- The `Holder` variable will be set when the weapon is attached to the player character or an enemy when the weapon is picked up. It is also used to determine whether the weapon spins when it is not picked up.

- The `Strength` variable is used to calculate the damage dealt to the hit target.

- The `IsWithinAttackRange` function will be implemented in *Chapter 9*.

This is the weapon.cpp code:

```
#include "Weapon.h"
#include "PlayerAvatar.h"
#include "DefenseTower.h"
```

```
AWeapon::AWeapon()
{
    PrimaryActorTick.bCanEverTick = true;

_StaticMesh =CreateDefaultSubobject<UStaticMeshComponent>
    (TEXT("Static Mesh"));SetRootComponent(_StaticMesh);

void AWeapon::BeginPlay()
{
Super::BeginPlay();

OnActorBeginOverlap.AddDynamic(this,
&AWeapon::OnWeaponBeginOverlap);
}

void AWeapon::Tick(float DeltaTime)
{
Super::Tick(DeltaTime);

  if (Holder == nullptr)
  {
    FQuat rotQuat = FQuat(
    FRotator(0, 300.0f * DeltaTime, 0));
    AddActorLocalRotation(rotQuat);
  }
  }
    void AWeapon::OnWeaponBeginOverlap(AActor* OverlappedActor,
    AActor* OtherActor)
  {
      auto character = Cast<ACharacter>(OtherActor);
  if (character == nullptr)
  {
    Return;
  }
  if (Holder == nullptr)
  {
    auto playerAvatar = Cast<APlayerAvatar>(character);
  if (playerAvatar != nullptr)
  {
    Holder = character;
    TArray<AActor*> attachedActors;
    OtherActor->GetAttachedActors(attachedActors, true);
```

```
    for (int i = 0;
    i < attachedActors.Num(); ++i)
  {
    attachedActors[i]-
    >DetachFromActor(FDetachmentTransformRules::KeepWorldTransform);
    attachedActors[i]->SetActorRotation(FQuat::Identity);
    AWeapon* weapon = Cast<AWeapon>(attachedActors[i]);
    weapon->Holder = nullptr;
  }
    AttachToComponent(Holder->GetMesh(),
    FAttachmentTransformRules::SnapToTargetIncludingScale,
    FName("hand_rSocket"));
  }
  }
  else if(IsWithinAttackRange(0.0f, OtherActor))
  {
    //deal damage to the target: hero or enemy
  }
}
```

Here is an explanation of the provided code:

- The `Tick()` function checks whether the `Holder` variable is `nullptr`. When the result is `true`, that means that the weapon has not been picked up, and the weapon spins to attract the player's attention.

- The `OnWeaponBeginOverlap` function starts by checking whether the overlapped `OtherActor` is a character. It casts the `OtherActor` pointer into a `character` pointer. If the casting failed, the `character` pointer value should be assigned `nullptr`.

- When the overlapped actor is a character, if the weapon's `Holder` is `nullptr`, it will attach to the player character if the overlapped character is `PlayerAvatar`. The `AttachToComponent` function is called to do the attachment work.

- Before attaching the weapon, we need to check and drop the player character's current weapon. Calling `OtherActor->GetAttachedActors()` can fill up the `attachedActors` array with all the actors (weapons, accessories, etc.) currently attached to the player character. We then use a `for` loop to detach all the found attached actors. In our case, the returned array length should always be 1 because the player character only picks up one weapon at a time. The `DetachFromActor` function is called to do the detachment work.

- Once a weapon is dropped, we call the actor's `SetActorRotation` function to set an identity quaternion, so that the weapon is placed straight on the ground.

- If the weapon has `Holder` when overlapping `OtherActor`, it should deal damage to the overlapped `OtherActor`. We will add the implementation code in *Chapter 9*.

- If `OtherActor` isn't a character, then we cast this `OtherActor` pointer to be a `DefenseTower` pointer. If the returned pointer is not `nullptr`, the weapon should deal damage to the overlapped tower. We will add the implementation code in *Chapter 9* too.

Play the game and move the hero through any weapon placed in the game level; you should see that the hero can pick up and drop weapons.

For the enemy, we can simply spawn a hammer and attach it to the enemy when the enemy begins to play. Let's see how to do this.

Spawning a weapon for the enemy

In *Pangaea*, we don't allow enemies to pick up weapons; instead, we spawn a particular weapon, the hammer, and attach it to its `Holder` enemy. To do that, we first need to add two protected variables, `WeaponClass` and `Weapon`, to the `AEnemy` class:

```
protected:
UClass* _WeaponClass;
AWeapon* _Weapon;
```

The `_WeaponClass` variable is used to store the blueprint class type value of the hammer. It is then used to instantiate the hammer.

Next, in the constructor of `AEnemy`, we can write the following two lines of code to find and set the `_WeaponClass` value:

```
static ConstructorHelpers::FObjectFinder<UBlueprint> blueprint_
finder(TEXT("Blueprint'/Game/TopDown/Blueprints/BP_Hammer.BP_
Hammer'"));

_WeaponClass = (UClass*)blueprint_finder.Object->GeneratedClass;
```

Here is an explanation of the code provided:

- The Unreal Engine's `ConstructorHelpers::FObjectFinder` struct helps find an asset from a given path in the project. We put `UBlueprint` between the two brackets as the template class, which indicates that the asset we want to find is a blueprint. The variable name is `blueprint_finder`.

- The asset path starts with the asset type specification, `Blueprint`, followed by an apostrophe (`'`). It implies that the actual path starts right after `Blueprint'`.

- The format of the blueprint asset name should be expressed as `<blueprintName>.<blueprintName>`. Here is an example – `BP_Hammer.BP_Hammer`.

- The returned value of `blueprint_finder.Object->GeneratedClass` is the found asset's `UClass` value, which can be assigned to `_WeaponClass`.

The task of finding and storing asset classes must be performed in the constructor of classes.

The next thing we want to do is to spawn the hammer in the `BeginPlay()` function:

```
_Weapon = Cast<AWeapon>(GetWorld()->SpawnActor(_WeaponClass));
_Weapon->Holder = this;
_Weapon->AttachToComponent(GetMesh(),
    FAttachmentTransformRules::SnapToTargetIncludingScale,
    FName("hand_rSocket"));
```

Here is an explanation of the code:

- The first line of this block of code calls the `SpawnActor()` function with `_WeaponClass` as the parameter to instantiate the hammer. The spawned actor pointer is cast as `AWeapon*` and assigned to the `_Weapon` variable.

- The second line assigns the `this` actor, which is the enemy itself, to be the holder of the weapon (the hammer).

- The third line attaches the hammer to `hand_rSocket` on the character's skeleton.

Now, the hero and the enemy have their weapons. Let's write some code to make the defense tower fire at an invading player.

Defense tower firing fireballs

Unlike the weapon, which has only one collision component, the defense tower has two collision components, `SphereComponent` and `StaticMeshComponent`:

- `SphereComponent` takes care of firing the `OnBeginComponentOverlap` and `OnEndComponentOverlap` events when the player character enters or leaves its scope, respectively.

- `StaticMeshComponent` acts like a collider that blocks pawns from moving through the tower. This collider is also used to generate `OnComponentHit` events when a player attacks the tower. Handling `OnComponentHit` will be described in *Chapter 9*.

Let's declare two more event handler functions for these two `SphereComponent` events in the `DefenseTower.h` header file:

```
UFUNCTION()
void OnBeginOverlap(UPrimitiveComponent* OverlappedComponent,
AActor* OtherActor,
UPrimitiveComponent* OtherComponent,
int32 OtherBodyIndex,
bool bFromSweep,
const FHitResult& SweepResult);
UFUNCTION()
void OnEndOverlap(UPrimitiveComponent* OverlappedComponent,
AActor* OtherActor,
UPrimitiveComponent* OtherComponent,
int32 OtherBodyIndex);
```

Both of the two functions have four common parameters that we are interested in here:

- The first parameter, `OverlappedComponent`, is the component that fires this event
- The second parameter, `OtherActor`, is the pointer to the other *entering or leaving* actor
- The third parameter, `OtherComponent`, is the other actor's component
- The fourth parameter, `OtherBodyIndex`, represents the other actor's body index

We also need a variable that holds the pointer to the target character. Since the target will only be the player, the pointer type can be `APlayerAvatar*`:

```
class APlayerAvatar* _Target = nullptr;
```

Note that we added the `class` keyword before `APlayerAvatar` here. The purpose of doing so is to not include the `PlayerAvatar.h` file in the `DefenseTower.h` header file, only in `DefenseTower.cpp`.

Using the `class` keyword here declares that `APlayerAvatar` is a class, so the compiler won't complain about the unknown symbol. When building `DefenseTower.cpp`, the compiler can still find the definition of the `APlayerAvatar` class because the `.cpp` file includes `PlayerAvatar.h` at the beginning of the file.

The next task that needs to be done is the implementation of the event handler functions. Handling the overlapping events in the event handler function is pretty straightforward. When the `OnComponentBeginOverlap` event is triggered, it simply sets `_Target` to `OtherActor`. Conversely, `_Target` is set to be `nullptr` when `OnComponentEndOverlap` is fired. We can see this code here:

```
void ADefenseTower::OnBeginOverlap(
    UPrimitiveComponent* OverlappedComponent,
    AActor* OtherActor,
    UPrimitiveComponent* OtherComponent,
    int32 OtherBodyIndex,
    bool bFromSweep,
    const FHitResult& SweepResult)
{
    APlayerAvatar* player = Cast<APlayerAvatar>(OtherActor);
    if (player)
    {
        _Target = player;
    }
}

void ADefenseTower::OnEndOverlap(
    UPrimitiveComponent* OverlappedComponent,
    AActor* OtherActor,
    UPrimitiveComponent* OtherComponent,
    int32 OtherBodyIndex)
{
    if (_Target != nullptr && OtherActor == _Target)
    {
        _Target = nullptr;
    }
}
```

The previously presented code demonstrates the implementation of the two event handler functions, OnBeginOverlap and OnEndOverlap. Now, the defense tower can sense the position of the player character within its attach range. The next step is to make the defense tower shoot fireballs at the player.

To fire fireballs, we need a UClass* variable to store the found fireball asset's class. Then, we can add the Fire() function to the ADefenseTower class. This function should be called in the Tick() function when _Target is a valid character. So, add the public function to DefenseTower.h:

```
void Fire();
```

In `DefenseTower.cpp`, implement the `Fire()` function and call it in the `Tick()` function:

```cpp
ADefenseTower::ADefenseTower()
{

    ...
  static ConstructorHelpers::FObjectFinder<UBlueprint> blueprint_
    finder(TEXT("Blueprint'/Game/TopDown/Blueprints/BP_Fireball.BP_
    Fireball'"));
    _FireballClass = (UClass*)blueprint_finder.Object->GeneratedClass;
}

void ADefenseTower::Fire()
{
  if(_Target == nullptr) return;
  auto fireball = Cast<AProjectile>(
  GetWorld()->SpawnActor(_FireballClass));
  FVector startLocation = GetActorLocation();
  startLocation.Z += 100.0f;
  FVector targetLocation = _Target->GetActorLocation();
  targetLocation.Z = startLocation.Z;
  FRotator rotation =
  UKismetMathLibrary::FindLookAtRotation(
  startLocation, targetLocation);
  fireball->SetActorLocation(startLocation);
  fireball->SetActorRotation(rotation);
}
void ADefenseTower::Tick(float DeltaTime)
{
  Super::Tick(DeltaTime);
  if (_Target != nullptr)
  {
    Fire();
  }
}
```

Let's break down this code:

- The `Fire()` function spawns the fireball from `_FireballClass` and places it at the location of the defense tower with a height of 100.0 units. `GetTargetLocation()` is the function that returns the actor's current location in an `FVector` struct.

- The fireball is also set with a rotation that makes the fireball orient to the current location of its target; this is because the fireball script will move the ball forward along with its orientation (see *Figure 8.17*). To set the fireball's rotation, we can use the location where the fireball is spawned as `startLocation`, get and use the target's current location as `targetLocation`, call the `FindLookAtRotation` function to get the rotator, and eventually, use the rotator as a parameter to invoke the `SetActorRotation` function.

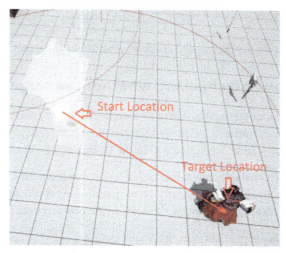

Figure 8.17 – Calculating the fireball rotation based on the start and target locations

- The `Tick()` function calls `Fire()` when there is a target to fire at.

Now, if you kick off the game and walk your hero close to a defense tower, the tower will fire hundreds of fireballs (see *Figure 8.18*). This problem is caused by the frequency of calling the `Fire()` function – the `Tick()` function is executed every frame, which means that it could be called 60 times every second when the frame rate is 60 fps.

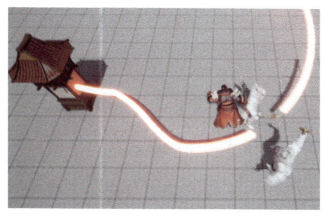

Figure 8.18 – The tower fires hundreds of fireballs in high frequently

Under this ad hoc case, we want to change the tower's tick time to be twice every second. That is 0.5 seconds for the tick interval. Calling `SetActorTickInterval()` can solve the problem, like so:

```
void ADefenseTower::BeginPlay()
{
    Super::BeginPlay();
    SetActorTickInterval(0.5f);
}
```

The last thing to do is to move the fireball and check whether it hits the target.

Moving the fireball and checking whether the target is hit

As a fireball moves forwards, it casts a ray ahead of its current position to check whether any obstacles are going to be hit in the next frame. If the fireball hits the player character, then it deals damage to the player. This functionality can be implanted within the `Projectile` class.

First, a fired fireball should constantly move forward in its set direction. This is done by getting the fireball's current location, calculating the velocity vector based on its speed and orientation, adding the velocity to the current location, and setting the actor to the new location, like so:

```
FVector currentLocation = GetActorLocation();
FVector vel = GetActorRotation().RotateVector(FVector::ForwardVector)
* Speed * DeltaTime;
    FVector nextLocation = currentLocation + vel;
    SetActorLocation(nextLocation);
```

To calculate the velocity (`vel`) vector, we use the fireball's current rotation to rotate the unit forward vector, which is actually `(1, 0, 0)`, and then we multiply the rotated vector by the speed and the delta time. The new location is the addition between the current location and the velocity.

Second, the fireball should have a limited lifespan. When the fireball travels for a certain number of seconds and still hits nothing, it should be destroyed by calling the `Destroy` function:

```
void AProjectile::BeginPlay()
{
    Super::BeginPlay();
    _LifeCountingDown = Lifespan;
}


void AProjectile::Tick(float DeltaTime)
{
    Super::Tick(DeltaTime);
  if ( _LifeCountingDown > 0.0f)
  {
```

```
...
    _LifeCountingDown -= DeltaTime;
}
else
{
    PrimaryActorTick.bCanEverTick = false;
    Destroy();
}
}
```

Here, the `_lifeCountingDown` variable is initialized in `BeginPlay()`. Then, the `_lifeCountingDown` variable is checked on every tick. If the value is greater than 0, reduce `DeltaTime` from it; otherwise, it will stop ticking and destroy the fireball.

The last and most important thing that the previous code does is detect whether it hits any targets while moving forward. It is possible to use the overlap events to check whether it hits any targets.

The `Projectile` class can be the parent of not only the fireball but also any other fired actors, such as a bullet. When a bullet moves too fast, it can skip the target without triggering any overlap events. *Figure 8.19* shows that when the fireball moves at a very fast speed, it may not hit the hero.

Figure 8.19 – A projectile moving fast may skip the player without firing overlap events

To avoid the aforementioned issue, a better solution is to cast a tracing ray in front of the projectile while it is moving. The length of the tracing ray can be set identically to the magnitude of the velocity. Then, we can choose to call an appropriate version of the `LineTrace` functions to detect any collisions.

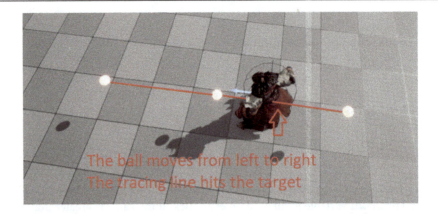

Figure 8.20 – A projectile moving fast with a tracing line can detect hitting the player

Unreal Engine offers a bunch of ray-tracing functions (`LineTrace` functions), including the following:

- Multiple-hit and single-hit line-tracing functions
- Synchronized and asynchronized line-tracing functions
- Tracing by channel, object type, and profile

You can visit the official documentation for the function details here: `https://docs.unrealengine.com/5.0/en-US/API/Runtime/Engine/Engine/UWorld/`.

In our case, we will use the `LineTraceSingleByObjectType()` function (the `LineTrace` function):

```
FHitResult hitResult;
FCollisionObjectQueryParams objCollisionQueryParams;
objCollisionQueryParams.AddObjectTypesToQuery(ECollisionChannel::ECC_
Pawn);

if(GetWorld()->LineTraceSingleByObjectType(hitResult,
currentLocation,
nextLocation,
objCollisionQueryParams))
{
auto playerAvatar = Cast<APlayerAvatar>(
hitResult.GetActor());
if (playerAvatar != nullptr)
{
playerAvatar->Hit(Damage);
PrimaryActorTick.bCanEverTick = false;
Destroy();
```

```
    }
  }
  ...
}
```

Here is an explanation of the code:

- We define a FHitResult type variable, hitResult, which will be used as the first parameter to call the LineTrace function.

- The second variable, objectCollisionQueryParams, is used as the fourth parameter to tell the LineTrace function what type of object it should collide with. The next line calls the AddObjectTypesToQuery method to add ECC_Pawn to the query parameter collection. This means that the LineTrace function will only check whether the tracing line hits any pawn target. This can be the player character or the enemy.

- The second and third parameters of the LineTrace function are the start and end locations of the tracing line.

- The LineTrace function returns true when a pawn is hit. The script checks whether the hit pawn is PlayerAvatar and deals damage when the check result is true.

Launch the game in the editor. You can use mouse clicks to walk around and pick up a weapon; when you are in a defense tower's attack range, the tower shoots fireballs at your character. The enemies will chase and attack you when your character is within their vision range; right-click on your mouse to fight back.

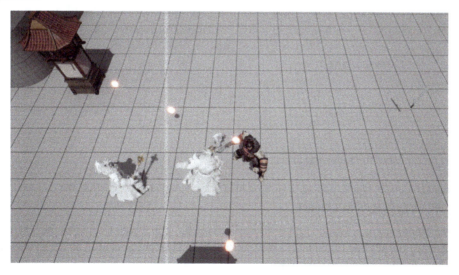

Figure 8.21 – Playing Pangaea in the editor

Processing a defense tower hit

In the *Defense tower firing fireballs* section, we associated two event handler functions with the actor's `BeginOverlap` and `EndOverlap` events. This configuration allows the tower to detect and fire at the player when the player's character enters its domain. However, to handle hits on the defense tower itself, we will adopt a different approach, which handles the `OnBeginOverlap` event of the tower's `MeshComponent`.

To initiate the implementation of this process, add a new `UFUNCTION` named `OnMeshBeginOverlap` at the end of `DefenseTower.h`:

```
UFUNCTION(BlueprintCallable)
void OnMeshBeginOverlap(AActor* OtherActor);
```

This function declaration indicates that this function can be called in Blueprint.

The function implementation should be added to `DefenseTower.cpp`:

```
void ADefenseTower::OnMeshBeginOverlap(AActor* OtherActor)
{
    AWeapon* weapon = Cast<AWeapon>(OtherActor);
    if (weapon == nullptr || weapon->Holder == nullptr)
    {
        return;
    }

    APangaeaCharacter* character = weapon->Holder;
    if (character->IsA(APlayerAvatar::StaticClass()) &&
        character->IsAttacking() &&
        CanBeDamaged())
    {
        Hit(weapon->Holder->Strength);
    }
}
```

This code accomplishes the following tasks:

- It casts the other overlapped actor to type `AWeapon*` and checks whether the casted actor is a valid weapon and has a holder
- If the other actor is a weapon and is held by a character, it proceeds to process the `Hit` event only when the holding character is a player avatar, the player avatar is in an attacking state, and the tower can currently be damaged
- The `Hit` function is called when all the previous conditions are met

With the `OnMeshBeginOverlap` function now prepared for invocation, let's proceed to add the `BeginOverlap` event node to the `BP_DefenseTower` event graph. This particular event node will handle the task of calling `OnMeshBeginOverlap` whenever the event is triggered.

Follow these steps to create the event node:

1. Open `BP_DefenseTower` in the editor.

2. Select **Mesh Component (Static Mesh)** in the **Components** panel.

Figure 8.22 – Selecting the Mesh Component in BP_DefenseTower

3. Click the + button to add the **Begin Overlap** node to the event graph.

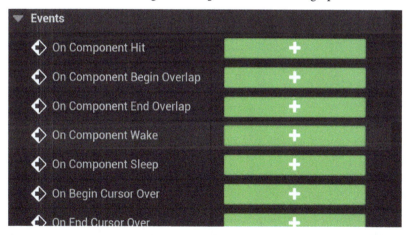

Figure 8.23 – Click the + button to add the BeginOverlap node to the event graph

4. Add the **On Mesh Begin Overlap** node.

5. Connect the execution pins from **On Component Begin Overlap** to **On Mesh Begin Overlap**.

6. Connect the **Other Actor** pin from **On Component Begin Overlap** to the pin of the **On Mesh Begin Overlap** node.

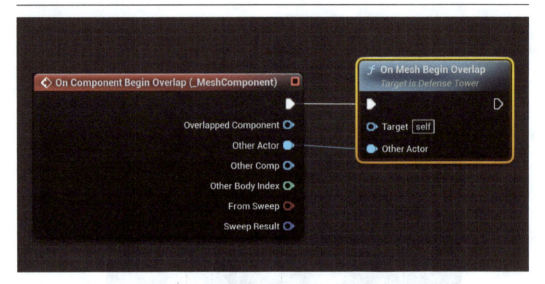

Figure 8.24 – Call the OnMeshBeginOverlap function when the
OnComponentBeginOverlap event is triggered

7.   Save and compile the BP_DefenseTower blueprint after your edits.

With the collision settings completed, the *Pangaea* game now possesses the essential interactions that make it enjoyable and playable.

Summary

In this chapter, you learned about collision detection and how to use the engine's collision system to bring interactions into the *Pangaea* game.

The first part of this chapter explained the three collision shape components (CapsuleComponent, BoxComponent, and SpherComponent) as well as the two mesh components (StaticMeshComponent and SekeltalMeshCbmponet). Adding collision shape components to actors and disabling the mesh collisions for better performance was also explained.

After that, you learned about the OnBeginOverlap and OnEndOverlap collision events and how they work with your C++ code. It was emphasized that you can handle collision events on either an actor or a component, depending on your actual needs.

Appropriate collision presets are vital to gameplay to properly process interactions between actors. It is important to understand what collision preset type, such as BlockAll, OverlapAllDynamic, and NoCollision, should be used for different collision components. You also learned how to customize a new collision preset for special uses.

Based on our knowledge of the mechanism of the collision system, we worked together to import and create the assets of weapons, the defense tower, and the fireball actors.

We coded a function to handle the weapon's `OnActorBeginOverlap` event to make the weapons pickable. We also added functions to handle the `OnComponentBeginOverlap` and `OnComponentEndOverlap` component events of `DefenseTower`. By doing that, the defense tower knew when the player was within its attack range and shot fireballs at the player character.

The last part of this chapter introduced the concept of ray casting and the engine's `LineTrace` functions. You learned to write the code to move the fireball and used the `LineTraceSingleByObjectType()` function to detect hitting the target.

The next chapter will focus on code refactoring and completing the rest of the gameplay (hit and damage processes, for example). Some useful UE5 APIs will be introduced here too.

9

Improving C++ Code Quality

Congratulations, you have completed eight chapters and have made the game playable! But do you notice something in the C++ code you wrote that makes you feel uncomfortable – for example, the duplicated variables and functions in the `APlayerAvatar` and `AEnemy` classes? This chapter will introduce two approaches (**code refactoring** and **code refinement**) that developers regularly use to improve their code quality.

Additionally, you will also learn how to output debug messages in Unreal and two ways (by calling the `Cast` and `IsA` functions) to find out an actor's class type.

By the end of this chapter, you will have a valuable awareness of the importance of maintaining code quality in programming, as well as knowledge regarding code refactoring, code refining, as well as the iterative process. This understanding will empower you to write high-quality and professional code.

Topics covered in this chapter include the following:

- Refactoring code
- Refining code
- Outputting debug messages
- Checking an `Actor` instance's actual class type

Technical requirements

The code for this chapter can be found at https://github.com/PacktPublishing/Unreal-Engine-5-Game-Development-with-C-Scripting/tree/main/Chapter09.

Refactoring code

You probably felt a little uncomfortable with the redundant code blocks we wrote in the previous chapters; for example, the `PlayerAvatar` and `Enemy` classes have some identical attribute variables and functions. Can we instead combine them and maintain only one copy of code for those replicated variables and functions? Yes, and this will require us to refactor the code.

Code refactoring helps improve the internal structure of code so that it becomes more readable, maintainable, and efficient. The process includes algorithm optimization, duplicate removal, and code simplification.

For the *Pangaea* project, we identified two refactoring tasks:

- Combining the two animation instance classes
- Adding a parent class for the `PlayerAvatar` and `Enemy` classes

Let's start with the first task.

Combining the PlayerAvatarAnimInstance and EnemyAnimInstance classes

Open and compare the `PlayerAvatarAnimInstance.h`, `PlayerAvatarAnimInstance.cpp`, `EnemyAnimInstance.h`, and `EnemyAnimInstance.cpp` files – you should find that the two pairs of header and source code files are almost identical, except for the class names.

For the *Pangaea* game, we know that the game design won't be changed in the future, so we can combine these two classes. We can create a new class called `PangaeaAnimInstance`, with the class files called `PangaeaAnimInstance.h` and `PangaeaAnimInstance.cpp`. Of course, the old files can be removed from the project (if you have forgotten how to remove code files, please refer to the *Recompiling C++ Projects* section of *Chapter 5*).

Here is the code for `PangaeaAnimInstance.h`:

```cpp
#pragma once
#include "CoreMinimal.h"
#include "Animation/AnimInstance.h"
#include "PangaeaAnimInstance.generated.h"

UENUM(BlueprintType)
enum class ECharacterState : uint8
{
    Locomotion,
    Attack,
    Hit,
    Die
```

```
};

UCLASS()
class PANGAEA_API UPangaeaAnimInstance : public UAnimInstance
{
   GENERATED_BODY()
public:
UPROPERTY(EditAnywhere, BlueprintReadWrite, Category =
"PangaeaAnimInstance Params")
   float Speed;
UPROPERTY(EditAnywhere, BlueprintReadWrite, Category =
"PangaeaAnimInstance Params")
   ECharacterState State;

   UFUNCTION(BlueprintCallable)
   void OnStateAnimationEnds();
};
```

Here is the code for PangaeaAnimInstance.cpp:

```
#include "PangaeaAnimInstance.h"
#include "PangaeaCharacter.h"

void UPangaeaAnimInstance::OnStateAnimationEnds()
{
if (State == ECharacterState::Attack)
{
   State = ECharacterState::Locomotion;
}
else
{
auto character = Cast<APangaeaCharacter>(GetOwningActor());

if (State == ECharacterState::Hit)
{
   if (character->GetHealthPoints() > 0.0f)
   {
     State = ECharacterState::Locomotion;
   }
   else
   {
     State = ECharacterState::Die;
   }
}
```

```
else if (State == ECharacterState::Die)
{
  character->DieProcess();
}
}
}
```

The previous code for the `PangaeaAnimInstance` class combined the redundant code in both `PlayerAvatarAnimInstance` and `EnemyAnimInstance` into one place. Since the two animation blueprints, `ABP_Player` and `ABP_Enemy`, were created based on the old animation instance classes, they both need to be reparented to the new `PangaeAnimInstance` class.

To do that, open `ABP_Player` in the **Animation Blueprint Editor** and select **File | Reparent Blueprint** from the main menu:

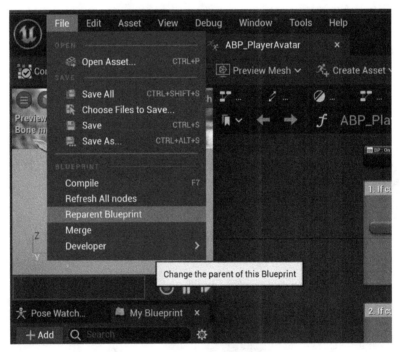

Figure 9.1 – Reparenting ABP_Player to PangaeaAnimInstance.png

Then, choose `PangaeaAnimInstance` from the **Reparent Blueprint** list.

Unfortunately, you have to re-set up the state machine, variables, and animations for the blueprint (follow the steps introduced in the *Creating the State Machine on ABP_PlayerAvatar* section of *Chapter 6* to do this).

Figure 9.2 illustrates the different class diagrams before and after the refactoring process:

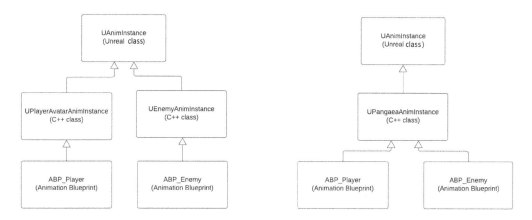

Figure 9.2 – Class diagrams before and after refactoring PlayerAvatarAnimInstance and EnemyAnimInstance into PangaeaAnimInstance

Next, we want to refactor the `PlayerAvatar` and `Enemy` classes to avoid having replicated attribute variables and functions.

Making PangaeaCharacter the parent class of APlayerAvatar and AEnemy

By investigating the `APlayerAvatar` and `AEnemy` classes, we can draw the class diagram, which includes information about the class variables and functions:

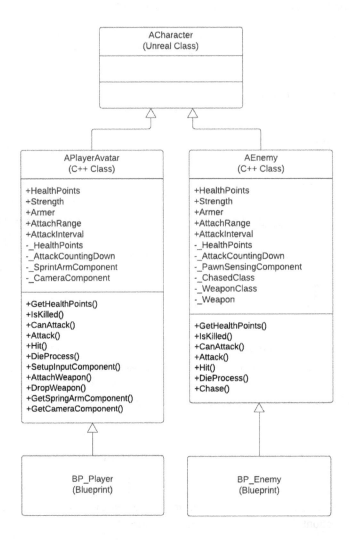

Figure 9.3 – Pangaea actors class diagram before the refactoring

In the diagram, + indicates that the subsequent variable or function is publicly accessible, and – indicates that the subsequent variable or function is only privately accessible.

As you can see, the APlayerAvatar class and the AEnemy class have some variables and functions in common, for instance, the HealthPoints variable and the GetHealthPoints function. The two copies of the same code may potentially cause inconsistency and confusion, which will make it harder to maintain the code. Therefore, we want to refactor these two classes and their relationship.

To refactor the class structure of `APlayerAvatar` and `AEnemy`, we can change and make the `APangaeaCharacter` class (which was already created with the *Pangaea* project) the parent class that is inherited by classes. Then, we can move the common variables and functions from the two child classes into the parent class:

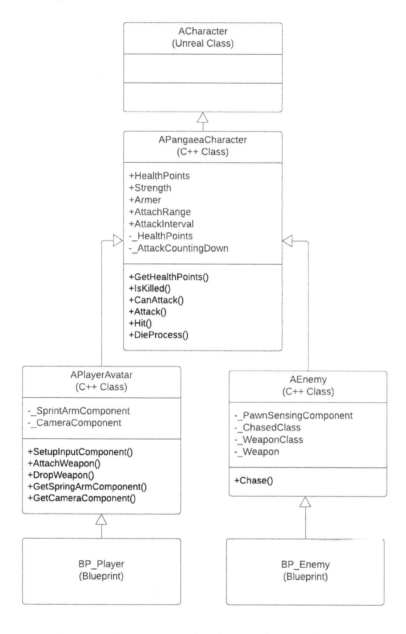

Figure 9.4 – Pangaea actors class diagram after the refactoring

As you can see, the common variables and functions are now written only once in the APangaea Character class, which helps avoid ambiguity and improves the code quality for maintenance.

We will present the code for the .h and .cpp files for the three refactored classes: APangaeaCharacter, APlayerAvatar, and AEnemy.

Here is the code for PangaeaCharacter.h:

```cpp
#pragma once

#include "CoreMinimal.h"
#include "GameFramework/Character.h"
#include "PangaeaCharacter.generated.h"

UCLASS(Blueprintable)
class APangaeaCharacter : public ACharacter
{
GENERATED_BODY()
public:
APangaeaCharacter();
UPROPERTY(EditAnywhere, Category = "Pangaea Character Params")
int HealthPoints = 100;

UPROPERTY(EditAnywhere, Category = "Pangaea Character Params")
float Strength = 5;

UPROPERTY(EditAnywhere, Category = "Pangaea Character Params")
float Armer = 1;

UPROPERTY(EditAnywhere, Category = "Pangaea Character Params")
float AttackRange = 200.0f;

UPROPERTY(EditAnywhere, Category = "Pangaea Character Params")
float AttackInterval = 3.0f;

public :
virtual void Tick(float DeltaTime) override;

UFUNCTION(BlueprintCallable,
          Category = "Pangaea|Character",
          meta = (DisplayName = "Get HP"))
int GetHealthPoints();
```

```
UFUNCTION(BlueprintCallable, Category = "Pangaea|Character")
bool IsKilled();

UFUNCTION(BlueprintCallable, Category = "Pangaea|Character")
bool CanAttack();

virtual void Attack();
virtual void Hit(int damage);
virtual void DieProcess();

protected:
virtual void BeginPlay() override;

class UPangaeaAnimInstance* _AnimInstance;
int _HealthPoints;
float _AttackCountingDown;
};
```

In `PangaeaCharacter.h`, you may have noticed the new `UPangaeaAnimInstance`-type variable, `_AnimInstance`, of the `APangaeaCharacter` class. This variable is used as a cache variable. We will explain it in the *Refining code* section.

We also changed the `APlayerAvatar` and `AEnemy` classes' parent to be `APangaeaCharacter` and removed those variables and functions that were already moved to the parent class.

The code for `PangaeaCharacter.cpp` is provided here:

```
#include "PangaeaCharacter.h"
#include "PangaeaAnimInstance.h"
#include "UObject/ConstructorHelpers.h"
#include "Camera/CameraComponent.h"
#include "Components/DecalComponent.h"
#include "Components/CapsuleComponent.h"
#include "GameFramework/CharacterMovementComponent.h"
#include "GameFramework/PlayerController.h"
#include "GameFramework/SpringArmComponent.h"
#include "Materials/Material.h"
#include "Engine/World.h"

APangaeaCharacter::APangaeaCharacter()
{
    PrimaryActorTick.bCanEverTick = true;
}
```

```cpp
void APangaeaCharacter::BeginPlay()
{
    Super::BeginPlay();

    _AnimInstance = Cast<UPangaeaAnimInstance>(
                            GetMesh()->GetAnimInstance());
    _HealthPoints = HealthPoints;
}

void APangaeaCharacter::Tick(float DeltaSeconds)
{
    Super::Tick(DeltaSeconds);
}

int APangaeaCharacter::GetHealthPoints()
{
    return _HealthPoints;
}

bool APangaeaCharacter::IsKilled()
{
    return (_HealthPoints <= 0.0f);
}

bool APangaeaCharacter::CanAttack()
{
    return (_AttackCountingDown <= 0.0f &&
     _AnimInstance->State == ECharacterState::Locomotion);
}

void APangaeaCharacter::Attack()
{
    _AttackCountingDown = AttackInterval;
}

void APangaeaCharacter::Hit(int damage)
{
    _HealthPoints -= damage;
    _AnimInstance->State = ECharacterState::Hit;

if (IsKilled()) {
        PrimaryActorTick.bCanEverTick = false;
}
```

```
}

void APangaeaCharacter::DieProcess()
{
    PrimaryActorTick.bCanEverTick = false;
    Destroy();
    GEngine->ForceGarbageCollection(true);
}
```

Next up is the code for `PlayerAvatar.h`:

```
#pragma once
#include "CoreMinimal.h"
#include "GameFramework/SpringArmComponent.h"
#include "Camera/CameraComponent.h"
#include "PangaeaCharacter.h"
#include "Weapon.h"
#include "PlayerAvatar.generated.h"

UCLASS(Blueprintable)
class PANGAEA_API APlayerAvatar : public APangaeaCharacter
{
GENERATED_BODY()
public:
APlayerAvatar();
protected:
virtual void BeginPlay() override;
public:
virtual void Tick(float DeltaTime) override;
virtual void SetupPlayerInputComponent(
        class UInputComponent* PlayerInputComponent) override;

UFUNCTION(BlueprintCallable,
                Category = "Pangaea|PlayerAvatar")
void AttachWeapon(AWeapon* Weapon);

UFUNCTION(BlueprintCallable,
                Category = "Pangaea|PlayerAvatar")
void DropWeapon();
void Attack() override;

FORCEINLINE
class UCameraComponent* GetCameraComponet() const
{ return _CameraComponent; }
```

```
FORCEINLINE
class USpringArmComponent* GetSringArmComponet() const { return _
SpringArmComponent; }

private:
UPROPERTY(VisibleAnywhere, BlueprintReadOnly,
  Category = "Camera", meta = (AllowPrivateAccess = "true"))
USpringArmComponent* _SpringArmComponent;

UPROPERTY(VisibleAnywhere, BlueprintReadOnly,
  Category = "Camera", meta = (AllowPrivateAccess = "true"))
UCameraComponent* _CameraComponent;
};
```

The implementation code for `PlayerAvatar.cpp` is provided here:

```
#include "PlayerAvatar.h"
#include "GameFramework/CharacterMovementComponent.h"
#include "PangaeaAnimInstance.h"

APlayerAvatar::APlayerAvatar()
{
bUseControllerRotationPitch = false;
bUseControllerRotationYaw = false;
bUseControllerRotationRoll = false;

auto characterMovement = GetCharacterMovement();
characterMovement->bOrientRotationToMovement = true;
characterMovement->RotationRate = FRotator(0.f, 640.f, 0.f);
characterMovement->bConstrainToPlane = true;
characterMovement->bSnapToPlaneAtStart = true;

_SpringArmComponent =
        CreateDefaultSubobject<USpringArmComponent>(
                            TEXT("SpringArm"));
_SpringArmComponent->SetupAttachment(RootComponent);
_SpringArmComponent->SetUsingAbsoluteRotation(true);
_SpringArmComponent->TargetArmLength = 800.f;
_SpringArmComponent->SetRelativeRotation(
                            FRotator(-60.f, 0.f, 0.f));
_SpringArmComponent->bDoCollisionTest = false;

CameraComponent =
```

```
        CreateDefaultSubobject<UCameraComponent>(TEXT("Camera"));
_CameraComponent->SetupAttachment(_SpringArmComponent,
                             USpringArmComponent::SocketName);
_CameraComponent->bUsePawnControlRotation = false;
}


void APlayerAvatar::BeginPlay()
{
Super::BeginPlay();
}


void APlayerAvatar::Tick(float DeltaTime)
{
Super::Tick(DeltaTime);

_AnimInstance->Speed =
            GetCharacterMovement()->Velocity.Size2D();

if (_AttackCountingDown == AttackInterval)
{
  _AnimInstance->State = ECharacterState::Attack;
}

if (_AttackCountingDown > 0.0f)
{
  _AttackCountingDown -= DeltaTime;
}
}


void APlayerAvatar::SetupPlayerInputComponent(
                    UInputComponent* PlayerInputComponent)
{
Super::SetupPlayerInputComponent(PlayerInputComponent);
}


void APlayerAvatar::AttachWeapon(AWeapon* Weapon)
{
Weapon->AttachToComponent(GetMesh(),
   FAttachmentTransformRules::SnapToTargetIncludingScale,
   FName("hand_rSocket"));
}


void APlayerAvatar::DropWeapon()
```

```
{
TArray<AActor*> attachedActors;
GetAttachedActors(attachedActors, true);
for (int i = 0; i < attachedActors.Num(); ++i)
{
  attachedActors[i]->DetachFromActor(
              FDetachmentTransformRules::KeepWorldTransform);
  attachedActors[i]->SetActorRotation(FQuat::Identity);
  AWeapon* weapon = Cast<AWeapon>(attachedActors[i]);
  if (weapon != nullptr)
  {
    weapon->Holder = nullptr;
  }
}
}

void APlayerAvatar::Attack()
{
  APangaeaCharacter::Attack();
}
```

Lastly, we'll modify the code for the AEnemy class. Here is the code for Enemy.h:

```
#pragma once

#include "CoreMinimal.h"
#include "PangaeaCharacter.h"
#include "Weapon.h"
#include "Enemy.generated.h"

UCLASS()
class PANGAEA_API AEnemy : public APangaeaCharacter
{
  GENERATED_BODY()

public:
  AEnemy();

protected:
  virtual void BeginPlay() override;

  APawn* _chasedTarget = nullptr;
  UClass* _WeaponClass;
  AWeapon* _Weapon;
```

```
public:
  virtual void Tick(float DeltaTime) override;

  void Attack() override;
  void DieProcess() override;

  UFUNCTION(BlueprintCallable, Category = "Pangaea|Enemy")
  void Chase(APawn* targetPawn);
private:
  UPROPERTY(VisibleAnywhere, BlueprintReadOnly,
                       meta = (AllowPrivateAccess = "true"))
  class UPawnSensingComponent* PawnSensingComponent;
};
```

The implementation code for Enemy.cpp is as follows:

```
#include "Enemy.h"
#include "Perception/PawnSensingComponent.h"
#include "GameFramework/CharacterMovementComponent.h"
#include "EnemyController.h"
#include "PangaeaAnimInstance.h"

AEnemy::AEnemy()
{
PawnSensingComponent =
  CreateDefaultSubobject<UPawnSensingComponent>(
                              TEXT("PawnSensor"));
static ConstructorHelpers::FObjectFinder<UBlueprint>
blueprint_finder(
TEXT("Blueprint'/Game/TopDown/Blueprints/BP_Hammer.BP_Hammer'"));
_WeaponClass = (UClass*)blueprint_finder.Object->GeneratedClass;
}


void AEnemy::BeginPlay()
{
Super::BeginPlay();

_Weapon = Cast<AWeapon>(GetWorld()->SpawnActor(_WeaponClass));
_Weapon->Holder = this;
_Weapon->AttachToComponent(GetMesh(),
     FAttachmentTransformRules::SnapToTargetIncludingScale,
     FName("hand_rSocket"));
}
```

```
void AEnemy::Tick(float DeltaTime)
{
Super::Tick(DeltaTime);

_AnimInstance->Speed =
       GetCharacterMovement()->Velocity.Size2D();

if (_AttackCountingDown == AttackInterval)
{
  _AnimInstance->State = ECharacterState::Attack;
}

if (_AttackCountingDown > 0.0f)
{
  _AttackCountingDown -= DeltaTime;
}

if (_chasedTarget != nullptr &&
    _AnimInstance->State == ECharacterState::Locomotion)
{
auto enemyController =
       Cast<AEnemyController>(GetController());
enemyController->MakeAttackDecision(_chasedTarget);
}
}

void AEnemy::Chase(APawn* targetPawn)
{
if (targetPawn != nullptr &&
    _AnimInstance->State == ECharacterState::Locomotion)
{
auto enemyController =
    Cast<AEnemyController>(GetController());
enemyController->MoveToActor(targetPawn, 90.0f);
}
_chasedTarget = targetPawn;
}

void AEnemy::DieProcess()
{
  Super::DieProcess();
  _Weapon->Destroy();
```

```
}

void AEnemy::Attack()
{
    APangaeaCharacter::Attack();

    GetController()->StopMovement();
}
```

The previously presented header and .cpp files demonstrated the code implementation for adding the `APangaeaCharacter` class and refactoring the `APlayerAvatar` and `AEnemy` classes. Moreover, we employed `_AnimInstance` as a cache variable to avoid constantly retrieving the character's animation instance.

In fact, using caching variables for improved performance is a process of refinement. Let's proceed to explain and carry out additional refinement tasks.

Refining code

Code refinement is performed to improve the overall quality of code and make changes to its external behavior. This usually involves improving performance, enhancing functionality, adding new features, and fixing bugs, which in turn improve the user experience.

We have identified two ways to refine the code:

- Using caching variables
- Creating a fireball pool

Let's go through each one.

Using caching variables

A caching variable is used to store a retrieved value that will be frequently used in the future. If you review the old `APlayerAvatar` and `AEnemy` source code, you will find the following two lines of code are executed at every frame in the `Tick` functions of the two classes:

```
auto animInst = Cast<UPlayerAvatarAnimInstance>(
                        GetMesh()->GetAnimInstance());

auto animInst = Cast<UEnemyAnimInstance>(
                        GetMesh()->GetAnimInstance());
```

If you look further into the implementations of the `CanAttack`, `Hit`, and `Chase` functions, the operation of getting and casting the character's animation instance is also executed.

We don't want to waste CPU time by repeatedly retrieving the animation instance all the time. Therefore, we use the caching variable, _AnimInstance, to store the retrieved pointer value in the APangaeaCharacter::BeginPlay function, which is called only once at the beginning; we can then directly read the value when needed:

```
void APangaeaCharacter::BeginPlay()
{

    ...

    _AnimInstance = Cast<UPangaeaAnimInstance>(
                        GetMesh()->GetAnimInstance());

    ...

}

bool APangaeaCharacter::CanAttack()
{
return (_AttackCountingDown <= 0.0f &&
        _AnimInstance->State == ECharacterState::Locomotion);
}
```

Having just introduced the utilization of caching variables as part of the code refinement process, our next objective is to implement a fireball pool. This pool will enhance the performance of managing pre-spawned fireballs and avoiding frequent spawns and despawns.

Creating a fireball pool

A fireball pool is created based on the concept of the **Object Pool** pattern, which allows the reuse of fireballs to avoid the overhead of creating and destroying them frequently. A fireball pool is essentially a collection of fireballs.

When shooting a fireball, the system checks and tries retrieving and reusing a fireball created upfront; otherwise, a new fireball is instantiated. Once a fireball's lifetime runs out or it hits something, instead of destroying it, we hide the fireball and add it to the fireball pool for future use.

The main benefits of using a fireball pool are as follows:

- **Better performance**: A fireball pool helps reduce the overhead involved in the frequent creation and destruction of fireballs

- **Reduction in the chances of facing the memory management problem**: By reusing fireballs, the number of memory allocations and garbage collections can be reduced, which avoids the problem of memory fragmentation

- **Control of the number of fireballs created in the system**: The fireball pool makes it possible to centralize the control over the maximum number of fireballs, which avoids the potential memory overflow problem

To create a fireball pool, we need a data container that will temporarily store inactive fireballs. Unreal provides different types of data containers, such as TArray, TQueue, and TMap (see *Table 9.1*), which we can choose from for the fireball pool:

Data container	Description
TArray	A data array that supports random access by index values. A loop must be used to search for an item by value.
TQueue	A data container that allows **First In, First Out** (**FIFO**) access operations. You can add items to the queue and dequeue items when it is not an empty queue.
TMap	A data container that stores pairs of keys and data values. It acts like a dictionary that allows querying values by keys. The search operation on TMap is based on a high-performance algorithm and is much faster than that on TArray.

Figure 9.5 – Descriptions of some Unreal data containers

For the *Pangaea* project, we'll simply use TQueue to create the FireballPool variable as a member of the APangaeaGameMode class. In this use case, the TQueue container is a suitable data structure that can satisfy the basic requirement. Making FireballPool a member of the game mode makes it convenient to globally access the fireball pool.

The following is the syntax of TQueue:

```
template<typename ItemType, EQueueMode Mode>
class TQueue
```

We also add two functions to APangaeaGameMode, the interfaces for retrieving and recycling fireballs:

```
AProjectile* SpawnOrGetFireball(UClass * ProjectileClass);
void RecycleFireball(AProjectile* projectile);
```

Let's break this code down:

- The SpawnOrGetProjectile function will spawn a new fireball actor if the fireball pool is empty; otherwise, it dequeues the first fireball from the pool and returns it
- The RecycleProjectile function simply enqueues the unused fireball onto the queue
- The parameter and return value of the fireball are of the AProjectile* type because AProjectile is the parent of the BP_Fireball blueprint

The new code for APangaeaGameMode.h is shown here:

```
#pragma once

#include "CoreMinimal.h"
```

```
#include "GameFramework/GameModeBase.h"
#include "Projectile.h"

#include "PangaeaGameMode.generated.h"

UCLASS(minimalapi)
class APangaeaGameMode : public AGameModeBase
{
GENERATED_BODY()

public:
APangaeaGameMode();
~APangaeaGameMode();

AProjectile* SpawnOrGetFireball(UClass * ProjectileClass);
void RecycleFireball(AProjectile* projectile);

protected:
TQueue<AProjectile*, EQueueMode::Spsc> _FireballPool;
};
```

The new code for `APangaeaGameMode.cpp` is shown here:

```
#include "PangaeaGameMode.h"
#include "PangaeaCharacter.h"
#include "PangaeaPlayerController.h"
#include "UObject/ConstructorHelpers.h"


APangaeaGameMode::APangaeaGameMode()
{
PlayerControllerClass = APangaeaPlayerController::StaticClass();

static ConstructorHelpers::FClassFinder<APawn>
PlayerPawnBPClass(TEXT("/Game/TopDown/Blueprints/BP_PlayerAvatar"));
if (PlayerPawnBPClass.Class != nullptr)
{
  DefaultPawnClass = PlayerPawnBPClass.Class;
}

static ConstructorHelpers::FClassFinder<APlayerController>
PlayerControllerBPClass(TEXT("/Game/TopDown/Blueprints/BP_
TopDownPlayerController"));
```

```
if(PlayerControllerBPClass.Class != NULL)
{
  PlayerControllerClass = PlayerControllerBPClass.Class;
}
}


APangaeaGameMode::~APangaeaGameMode()
{
AProjectile* fireball;
while (!_FireballPool.IsEmpty() &&
  _FireballPool.Dequeue(fireball))
{
  fireball->Destroy();
}
_FireballPool.Empty();
}

AProjectile* APangaeaGameMode::SpawnOrGetFireball(UClass*
projectileClass)
{
AProjectile* fireball = nullptr;

if (_FireballPool.IsEmpty())
{
  fireball = Cast<AProjectile>(
  GetWorld()->SpawnActor(projectileClass));
}
else
{
  _FireballPool.Dequeue(fireball);
  fireball->Reset();
}
return fireball;
}

void APangaeaGameMode::RecycleFireball(AProjectile* projectile)
{
  if (projectile == nullptr)
  {
    return;
  }

  projectile->SetActorHiddenInGame(true);
```

```
    projectile->SetActorEnableCollision(false);
    projectile->SetActorTickEnabled(false);
    _FireballPool.Enqueue(projectile);
}
```

For better comprehension, let's clarify the code:

- _FireballPool is defined as being of the TQueue type. The template specification tells Unreal that the queue elements should be of the AProjectile* type.

- When calling the RecycleFireball function to recycle a fireball, we use three steps (hide, disable collision, and disable ticking) to deactivate the fireball:

```
    projectile->SetActorHiddenInGame(true);
    projectile->SetActorEnableCollision(false);
    projectile->SetActorTickEnabled(false);
```

- When calling the SpawnOrGetFireball function, a fireball is dequeued from the pool; we call the fireball's Reset function to reactivate it.

Now, FireballPool is ready to be used. Let's make some changes to the DefenseTower and Projectile classes so that we can utilize FireballPool and improve the performance of the game.

To use the fireball pool, we need to make three changes to the ADefenseTower class.

First, we add a new member variable to _PangaeaGameMode:

```
class PANGAEA_API ADefenseTower : public AActor
{
    ...
protected:
    class APangaeaGameMode* _PangaeaGameMode;
    ...
};
```

Next, we retrieve and store the APangaeaGameMode instance to the _PangaeaGameMode variable in the ADefenseTower::BeginPlay function:

```
void ADefenseTower::BeginPlay()
{
    ...
    _PangaeaGameMode = Cast<APangaeaGameMode>(
    UGameplayStatics::GetGameMode(GetWorld()));
}
```

Finally, we call the `APangaeaGameMode::SpawnOrGetFireball` function instead of the `UWorld::SpawnActor` function to fire a fireball:

```
...
void ADefenseTower::Fire()
{
/* This block is commented
auto fireball = Cast<AProjectile>(
  GetWorld()->SpawnActor(_FireballClass));
*/
auto fireball = _PangaeaGameMode->SpawnOrGetFireball(
  _FireballClass);

  ...
}
```

Fireballs are deactivated, meaning they are hidden, and their collision and ticking functionalities are disabled when they are recycled to the pool; we need to add and use a new `Reset` function to the `AProjectile` class, which takes care of re-activating the fireball before reusing it.

We also want to add the `_PangaeaGameMode` variable to the `AProjectile` class to cache the `APangaeaGameMode` instance. The following code shows the changes in the `Projectile.h` file:

```
...
UCLASS(Blueprintable)
class PANGAEA_API AProjectile : public AActor
{
...
class APangaeaGameMode* _PangaeaGameMode;

public:
...
void Reset();
};
```

The provided code simply introduces a new variable, which is a pointer to the `APangaeaGame Mode` instance.

In `Projectile.cpp`, the primary objective of making this change is to recycle unused fireballs in the pool instead of destroying them:

```
...
void AProjectile::BeginPlay()
{
```

```cpp
Super::BeginPlay();
_PangaeaGameMode = Cast<APangaeaGameMode>(
      UGameplayStatics::GetGameMode(GetWorld()));
Reset();
}


void AProjectile::Tick(float DeltaTime)
{
Super::Tick(DeltaTime);

if (_LifeCountingDown > 0.0f)
{
  FVector currentLocation = GetActorLocation();
  FVector vel = GetActorRotation().RotateVector
    (FVector::ForwardVector) * Speed * DeltaTime;
  FVector nextLocation = currentLocation + vel;
  SetActorLocation(nextLocation);

  //Ray cast check
  FHitResult hitResult;
  FCollisionObjectQueryParams objCollisionQueryParams;
  objCollisionQueryParams.AddObjectTypesToQuery(
    ECollisionChannel::ECC_Pawn);

  if (GetWorld()->LineTraceSingleByObjectType(
    hitResult, currentLocation, nextLocation,
    objCollisionQueryParams))
  {
      auto playerAvatar = Cast<APlayerAvatar>(
      hitResult.GetActor());
    if (playerAvatar != nullptr)
    {
      playerAvatar->Hit(Damage);
        //Destroy();
      _PangaeaGameMode->RecycleFireball(this);
    }
  }

  //Reduce time
  _LifeCountingDown -= DeltaTime;
}
else
{
  //Destroy();
```

```
    _PangaeaGameMode->RecycleFireball(this);
  }
}

void AProjectile::Reset()
{
  _LifeCountingDown = Lifespan;
  SetActorHiddenInGame(false);
  SetActorEnableCollision(true);
  SetActorTickEnabled(true);
}
```

Here is the explanation of the primary tasks performed by the code:

- In `BeginPlay`, the `UGameplayStatics::GetGameMode` function gets and stores the `APangaeaGameMode` instance to the `_PangaeaGameMode` variable

- In `BeginPlay`, the `Reset` function is called to reset the projectile's states

- In `Tick`, `RecycleFireball` is called instead of calling the `Destroy()` function when the projectile hits something or its lifetime runs out

- The implementation of the `Reset()` function restores the projectile's lifespan, makes the projectile visible in the game, and enables the collision detection and the ticking functionality for the projectile

In addition to the previous processes, you can try to identify and fix some other bugs and add or remove code to make the gameplay smoother. Please download the source code files from the `Chapter09` folder in this book's GitHub repository and compare them with those in the `Chapter08` folder.

Refactoring and refining code is essential for programmers to develop good coding habits as well as produce efficient, maintainable, and adaptable software. Regularly refactoring and refining code helps identify and eliminate code smells, such as duplication, design defects, excessive complex code blocks, and potential bug causes, and generate robust and high-quality code.

After learning the skills of refactoring and refining code, we want to introduce another useful programming tool used to output log information during runtime.

Outputting debug messages

Using logs in programming has several benefits. Firstly, it can help developers to trace runtime information and use the clues to troubleshoot problems. Secondly, developers can also use logs to track and fix dynamic runtime errors that are usually difficult to identify during static debugging. Finally, logs can be used to monitor program running states for analyzing and improving performance, security, and quality. Logs are an essential tool for developers.

Here, we will introduce two ways through which you can output debug messages in Unreal: the UE_LOG macro and the AddOnScreenDebugMessage function.

Using the UE_LOG macro

UE_LOG is a macro that logs messages to either the **Output Log** window, the **Console**, or the log files. Use the tilde key (~) to open and close the console.

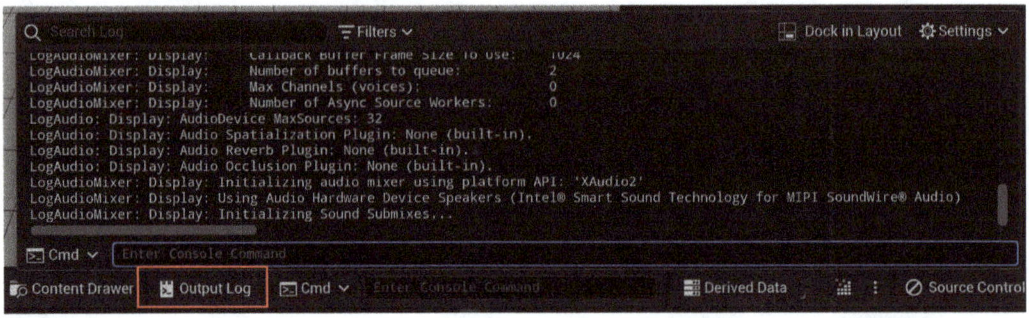

Figure 9.6 – UE5 editor Output Log window

The syntax of UE_LOG is as follows:

```
UE_LOG(<LogType>, <LogLevel>, <Message>)
```

The following is a description of the UE_LOG macro's parameters:

- LogType is the logging category name. Commonly used options are LogTemp, LogBlueprintUserMessage, and LogWorld.

- Verbosity Level tells the engine what verbosity level (log, warning, or error) message should be displayed and where the message should be output (the **Console**, the **Output Log** window, and/or the log files). Commonly used options are as follows:

  - Log: Prints a log message to the **Output Log** window and the log files. The color of the text of the message in the **Output Log** window is gray.

  - Warning: Prints a warning message to the **Output Log** window and the log files. The color of the text of the message in the **Output Log** window is yellow.

  - Error: Prints an error message to the **Output Log** window and the log files. The color of the text of the message in the **Output Log** window is red.

- Message is the actual text message. You can use the TEXT macro to format and convert C++ strings (the char* type) into FText type values, which will then be used as the third parameter of the UE_LOG macro.

All the log files are saved in subfolders of the /Saved folder under your game project folder.

The following example demonstrates how to output a message showing the Weapon overlapped! message when the OnWeaponBeginOverlap event is triggered:

```
void AWeapon::OnWeaponBeginOverlap(AActor* OverlappedActor, AActor*
OtherActor)
{
  UE_LOG(LogTemp, Log, TEXT("Weapon overlapped!"));
  ...
}
```

We just introduced the UE_LOG macro, which provides a way to print log information out to the **Output Log** window, the game console, and the log file. Let's now look at the AddOnScreenDebugMessage function to directly print debug messages onto the game screen.

Printing debug messages to the screen

Logging messages to the **Output Log** window, the **Console**, and the log files is useful, but it is more convenient to directly display messages on screen when you want to get some real-time debug information on a standalone game, for example, running and testing a game on mobile or VR devices.

For this, we can call the AddOnScreenDebugMessage function anytime in our program to display messages on the game screen. The syntax of this function is as follows:

```
void AddOnScreenDebugMessage
(
    uint64 Key,
    float TimeToDisplay,
    FColor DisplayColor,
    const FString & Message,
    bool NewerOnTop = true,
    const FVector2D & TextScale = FVector2D::One
)
```

Breaking this code down, we have the following:

- Key: A unique key for the message to prevent it from being added multiple times.
- TimeToDisplay: The number of seconds the message will stay on display.
- DisplayColor: The text color used for the message.
- Message: The actual text message.
- NewerOnTop: The default value is true, which indicates that new messages will be displayed before older messages.

- `TextScale`: The width and height scales of the message text. The default values are FVector2D::One (1.0, 1.0).

Here is an example to display the same message that UE_LOG output on the game screen:

```
void AWeapon::OnWeaponBeginOverlap(AActor* OverlappedActor, AActor*
OtherActor)
{
    GEngine->AddOnScreenDebugMessage(-1, 1.0f,
            FColor::Orange, TEXT("Weapon overlapped"));
    …
}
```

Please be aware that AddOnScreenDebugMessage is a member function of GEngine, which is a unique instance of the UEngine class.

After learning how to output debug messages, the last thing we want to introduce in this chapter is how to use the Cast and IsA functions to find out an Actor's actual class type. In addition, we will clarify the distinction and connection between the two functions, enabling you to select the more suitable function under any particular circumstances.

Checking an Actor instance's actual class type

In Unreal, the OOP approach regularly utilizes **inheritance** to declare class relations. Even though inheritance has the advantage of code reuse, it has the limitation of storing and returning instances while only knowing the base class types.

Based on the inheritance design pattern, type casting very often occurs in OOP. An example in Unreal scripting is that when a collision triggers an overlap event, the two parameters of the event function are both AActor* type pointers, whereas they should be of the APlayerAvatar*, AEnemy*, or ADefenseTower* type.

To find out whether AActor* is another child class type pointer, we use the Cast function to cast the AActor* pointer to the required type of pointer (APangaeaCharacter*, for example) and check whether the result is nullptr or not. If the Cast operation fails, it means that the actor is not the type of actor we need:

```
void AWeapon::OnWeaponBeginOverlap(AActor* OverlappedActor, AActor*
OtherActor)
{
    auto character = Cast<APangaeaCharacter>(OtherActor);
    if (character != nullptr)
    { … }
    else
```

```
{ ... }
}
```

This example shows how we check whether the other overlapped actor is a PangaeaCharacter-type actor.

Another way to check actor types is to call AActor's IsA function. The following code does the same job as the previous example by using the IsA function:

```
void AWeapon::OnWeaponBeginOverlap(AActor* OverlappedActor, AActor*
OtherActor)
{
  if (character->IsA(APangaeaCharacter::StaticClass()))
  { ... }
else
{ ... }
}
```

The question you may have is which of the functions, Cast or IsA, is better? The former does a little bit more work than the latter. The Cast function checks whether the input pointer is nullptr first, and then calls the IsA function to check and return the type. Therefore, the former is safer while the latter has better performance. Another benefit of using the Cast function is that it returns an expected subclass type pointer.

We put the full implementation for the AWeapon::OnWeaponBeginOverlap event function in the following code snippet, so that you can see how we use the two methods to check the collided actors' types:

```
void AWeapon::OnWeaponBeginOverlap(AActor* OverlappedActor, AActor*
OtherActor)
{
//GEngine->AddOnScreenDebugMessage(-1, 1.0f,
          FColor::Orange, TEXT("Weapon overlapped"));
//UE_LOG(LogTemp, Log, TEXT("Weapon overlapped"));

auto character = Cast<APangaeaCharacter>(OtherActor);
if (character != nullptr)
{
  if (Holder == nullptr)
  {
    auto playerAvatar = Cast<APlayerAvatar>(character);
    if (playerAvatar != nullptr)
    {
      Holder = character;
      playerAvatar->DropWeapon();
```

```
            playerAvatar->AttachWeapon(this);
        }
    }
    else if(character != Holder &&
        IsWithinAttackRange(0.0f, OtherActor) &&
        character->CanBeDamaged() &&
        Holder->IsAttacking())
    {
      character->Hit(Holder->Strength);
      if (character->IsA(APlayerAvatar::StaticClass()))
      {
        GEngine->AddOnScreenDebugMessage(- 1, 1.0f, FColor::Red,
          TEXT("Hit PlayerAvatar"));
        UE_LOG(LogTemp, Log, TEXT("Hit PlayerAvatar"));
      }
      else
      {
        GEngine->AddOnScreenDebugMessage(-1, 1.0f, FColor::Cyan,
          TEXT("Hit Enemy"));
        UE_LOG(LogTemp, Log, TEXT("Hit Enemy"));
      }
    }
}
else if(Holder != nullptr &&
        Holder->IsA(APangaeaCharacter::StaticClass()) &&
        Holder->IsAttacking())
{
  auto tower = Cast<ADefenseTower>(OtherActor);
  if (tower != nullptr &&
          tower->CanBeDamaged() &&
          IsWithinAttackRange(0.0f, tower))
  {
    tower->Hit(Strength);
    GEngine->AddOnScreenDebugMessage(-1, 1.0f, FColor::Cyan,
      TEXT("Hit Tower"));
  }
}
}
```

In the preceding code, we also completed the code to process hits to characters and towers by calling their Hit member functions. Please read the code and understand how the conditions are checked for the different hit processes.

Summary

In this chapter, you mainly learned how to improve the quality of the code in the *Pangaea* project, as well as how to output debug messages to monitor real-time game information.

Firstly, we introduced the concepts of code refactoring and code refinement, which made you aware of the importance of regular code quality improvement.

After that, we analyzed the *Pangaea* source code and identified two main issues that could be overcome: the redundant two animation instance classes of `UPlayerAnimInstance` and `UEnemyAnimInstance` and the duplicated member variables and functions of `APlayerAvatar` and `AEnemy`.

To resolve the first issue, you added a new class, `UPangaeaAnimInstance`, to replace the existing `UPlayerAnimInstance` and `UEnemyAnimInstance` classes. To resolve the second issue, you made the `APangaeaCharacter` class the parent class of `APlayerAvatar` and `AEnemy`, so that the duplicated variables and functions could be moved into the parent class.

Then, improving the code performance through code refinement, you used the `_AnimInstance` cache variable in the `APangaeaCharacter` class and added a fireball pool to avoid the frequent spawning and destroying of fireballs.

Finally, you learned how to use the `UE_LOG` macro and `AddOnScreenDebugMessage` to output debug messages, as well as using the `IsA` function to check whether an actor is of a certain class type.

In the next chapter, we will introduce some basic multiplayer game elements and demonstrate how to make *Pangaea* a multiplayer game.

Part 3: Making a Complete Multiplayer Game

In this part, we will delve into Unreal Engine's multiplayer support and guide you through writing C++ code to convert some *Pangaea* game actors into networked actors, explaining the fundamental concepts of client/server and the multiplayer modes.

To create a comprehensive multiplayer game, you will also learn how to extend the games' core classes, such as `GameMode`, `GameState`, and `GameInstance`, as well as create the multiplayer game's menu and the HUD for the game flow, so that players can start the server or join a session to play the game.

Additionally, you will learn valuable techniques for optimizing the game and utilizing high-quality assets to enhance its polish. Plus, we will cover the process of packaging the game for distribution.

This part contains the following chapters:

- *Chapter 10, Making Pangaea a Network Multiplayer Game*
- *Chapter 11, Controlling the Game Flow*
- *Chapter 12, Polishing and Packaging the Game*

10

Making Pangaea a Network Multiplayer Game

Unreal Engine was initially developed for the *Unreal* **first-person shooter** (**FPS**) game in 1998, and the game's multiplayer mode allowed up to 16 players to join a session. Owing to the game's popularity, Epic Games released *Unreal Tournament* in 1999, and it was one of the most popular multiplayer shooter games.

Multiplayer support is one of the key strengths of Unreal Engine; it is, in fact, one of the reasons why the engine is so popular and successful. This chapter will walk you through the steps to convert *Pangaea* into a multiplayer game so that you can learn how to start the game as a server or a client, synchronize game states, and send messages between the server and connected clients.

We will cover the following topics in this chapter:

- Comparing single-player and multiplayer games
- Launching the multiplayer *Pangaea* game in the editor
- Understanding multiplayer game network modes
- Handling network synchronizations

Technical requirements

The code for this chapter can be found at `https://github.com/PacktPublishing/Unreal-Engine-5-Game-Development-with-C-Scripting/tree/main/Chapter10`.

Comparing single-player and multiplayer games

A single-player game runs on a local machine (PC, game console, and so on) as a standalone application. The player interacts with the game directly through one or more connected input devices (keyboard, mouse, game pad, and so on). The game simulation is handled on the local machine. Here's how its setup looks:

Figure 10.1 – Single-player game

As with single-player games, a local multiplayer game runs on a local machine as a standalone application. The game allows multiple players (usually one to four players) to play the game together and interact with the game directly through their input devices. The game simulation is still handled on the local machine. The setup looks like this:

Figure 10.2 – Local multiplayer game

A network multiplayer game allows multiple players to play together on different types of devices (PCs, game consoles, mobile devices, and so on) over a network connection. A network multiplayer game usually involves a server and connected game clients. Players interact with each other in real time on the game clients, and their inputs are sent to the server for the game simulation. The game states are then synced to clients for the displays. You can see an example setup of a network multiplayer game here:

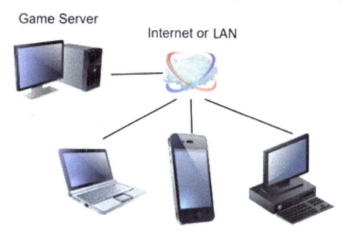

Figure 10.3 – Network multiplayer game

You should now have a basic idea of how network multiplayer games differ from single-player and local multiplayer games. But you may wonder: *does Unreal Engine support network multiplayer game development?* The answer is *yes*. Let's launch and play the multiplayer game in the engine.

Launching the multiplayer Pangaea game in the editor

The first important thing you need to know is that Unreal Engine is internally designed based on the network multiplayer mechanism, which means that all games built with Unreal, irrespective of whether they are single-player or multiplayer, are **client/server (CS)** multiplayer games.

To verify whether our *Pangaea* game supports multiplayer mode, you can launch the game with two players in the Unreal Editor. To do that, click the **Change Play Mode and Play Settings** button from the toolbar and set **Number of Players** to **2**:

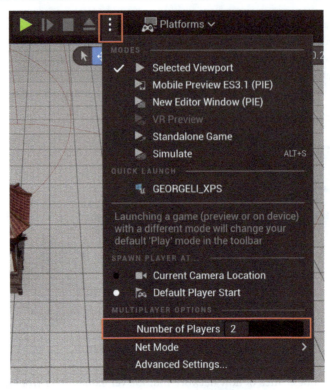

Figure 10.4 – Setting the number of players

Then, select **Net Mode** from the same drop-down menu and select **Play As Listen Server**:

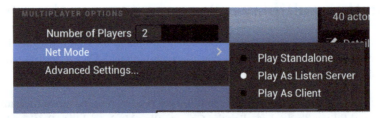

Figure 10.5 – Setting the play mode as Play as Listen Server

Now, click the **Play** button (the green triangle) to launch the game. You should get two game windows (one in the viewport and another one as a separate window), and you will notice that the game spawned two heroes in the scene (see *Figure 10.6*):

Figure 10.6 – Launching Pangaea with two players

If you click to set the focus on the separate game window and try moving your hero around, you will see that the hero moves in the other viewport game window. Picking up weapons and enemy moves are also synced between the two game windows.

This way, we just launched the game, then started a server and a client. The game in the editor's viewport was acting as a listen server (a host), whereas the other game in a separate window was a pure client.

Let's learn and understand more about multiplayer game network modes.

Understanding multiplayer game network modes

Network mode refers to a game application's relationship to a multiplayer game session. A game instance can adopt any of the following game modes:

- **Listen server (host)**: In this mode, one player's machine serves as both the game server and the local player's game environment. The game running on the server is hosting a network multiplayer session, which accepts connections from remote clients. This mode is often used for casual cooperative and competitive multiplayer games.

- **Dedicated server**: In this mode, the game is running as a server hosting a network multiplayer session. It accepts connections from remote clients but has no local players. The dedicated server mode is often used for large-scale multiplayer games.

A headless dedicated server is a server packaged without any graphic and audio, input, or other player-oriented features, so that the server runs more efficiently and uses less memory.

- **Client**: In this mode, the game is running as a client that is connected to a server. Player inputs are accepted on the clients and sent to the server for game simulation. Game states are synced from the server to the clients for visualizing the game scene to the players.

An Unreal game application can run as either a listen server, a dedicated server, or a client. We will introduce how to write code to execute the console command to start the game as a listen server or a client in *Chapter 12*.

Next, let's delve into the process of synchronizing actors across the client-server gameplay network.

Handling network synchronizations

When you played the multiplayer *Pangaea* game, you would have noticed some bugs; for example, if you attacked on the client side, the player's server avatar wouldn't attack. These bugs occurred because our single-player code didn't handle multiplayer synchronizations.

Before writing multiplayer code, we want to emphasize that you must be very clear in your mind whether the code will be executed on the server side, the client side, or both the server and client sides.

To make *Pangaea* playable as a multiplayer game, we need to do the following:

- Notify player attacks with **remote procedure calls (RPCs)**
- Sync actor variables to clients with **replications**
- Update the character health bar with `RepNotify`
- Process hits on the server
- Spawn fireballs on the server side

Let's look at how we can implement all of these listed actions.

Notifying player attacks with RPCs

In *Pangaea*, a player's attack events are triggered when the player clicks the right mouse button on the client side. This event message should be sent to the server so that the server can process the hit and broadcast the hit and the result to all clients.

We can use RPCs to complete this work. An RPC is a mechanism that allows a program running on one device to invoke a procedure on another device over a network.

A remote procedure can be called from any remote machine and executed on a specific machine that is connected to the same network session.

There are three types of RPCs, as listed here:

- **Server** RPCs are called from clients and executed on the server
- **Client** RPCs are usually called from the server and executed on a client that owns the receiving actor
- **NetMulticast** RPCs are called on the server and executed on all clients as well as the server itself

RPCs can be either **reliable** or **unreliable**. Tagging an RPC as `reliable` or `unreliable` indicates whether the RPC function is guaranteed to arrive at its intended destination. Multiplayer games usually send critical messages (attack and chat messages, for example) by calling reliable RPCs. For non-critical messages (move to the next location, for example), unreliable RPCs can be used to achieve faster networking performance, but they introduce the risk of message loss.

RPC functions should be tagged with the `UFUNCTION` macro, which comes with one of the RPC-type specifiers (`Server`, `Client`, or `NetMulticast`) and, optionally, a reliability specifier (`reliable` or `unreliable`). RPC functions without a reliability specifier are by default reliable RPC functions.

To solve the problem of out-of-sync attack animation for *Pangaea*, we can add the following RPC functions:

- On the client side, once the player inputs an attack command, `APlayerAvatar` should call the **Server** RPC (`Attack_RPC()`) to notify the server of this action
- On the server side, once an attack message is received, `APangaeaCharacter` calls the **NetMulticast** RPC (`Attack_Broadcast_RPC()`) to notify all clients

In *Figure 10.7*, you can observe how the RPC functions are called to synchronize the server and the clients during gameplay, using the attack use case:

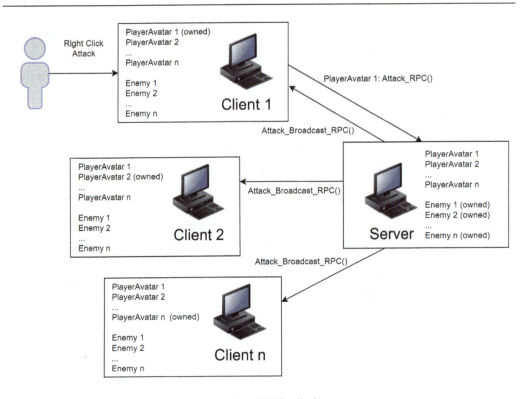

Figure 10.7 – Attack RPC calls diagram

The preceding diagram demonstrates how a player's input is handled and synchronized across the network server and clients, as follows:

- The player in **Client 1** attacks by clicking the right mouse button

- **Client 1** calls `Attack_RPC()` from the server

- The server processes the attack and calls `Attack_Broadcast_RPC()` to notify all clients, and all the clients play `PlayerAvatar 1`'s attack animation

You may be wondering why the `Attack_Broadcast_RPC()` function is a member of `APangaeaCharacter` instead of `APlayerAvatar`. This is because `APangaeaCharacter` is a parent class, so the RPC function can be shared by both `APangaea` and `AEnemy` to broadcast **Attack** events.

To add the `Attack_RPC()` function to `APlayerAvatar` and the `Attack_Broadcast_RPC()` function to `APangaeaCharacter`, we can start by declaring the two RPC functions for `APlayerAvatar` and `APangaeaCharacter`.

First, append the declaration of the `Attack_RPC()` function in `PlayerAvatar.h`, like so:

```
UFUNCTION(Server, Reliable)
void Attack_RPC();
```

Second, add the declaration of the `Attack_Broadcast_RPC()` function in `PangaeaCharacter.h`, as follows:

```
UFUNCTION(NetMultiCast, Reliable)
void Attack_Broadcast_RPC();
```

Both of the two new RPC functions are tagged with the `reliable` specifier, telling the engine that the RPC messages must be guaranteed to reach their destinations without fail.

Third, the implementation of `Attack_RPC()` should be added to the `PlayerAvatar.cpp` file, as follows:

```
void APlayerAvatar::Attack_RPC_Implementation()
{
    Attack_Broadcast_RPC();
}
```

Finally, the implementation of `Attack_Broadcast_RPC()` should be added to the `PangaeaCharacter.cpp` file, like so:

```
void APangaeaCharacter::Attack_Broadcast_RPC_Implementation()
{
    Attack();
}
```

> **Note**
>
> The implemented RPC functions must be appended with the `_Implementation` suffix.

Now, we can slightly change `APangaeaPlayerController::OnAttackPressed()` to call `Attack_RPC()` when the **Attack** button is clicked on the client side. So, add the following code to `PangaeaPlayerController.cpp`:

```
void APangaeaPlayerController::OnAttackPressed()
{
auto playerAvatar = Cast<APlayerAvatar>(GetPawn());
if (playerAvatar != nullptr && playerAvatar->CanAttack())
{
StopMovement();
```

```
playerAvatar->Attack_RPC();
    }
}
```

Using RPCs to sync an attack action between servers and clients has limitations. For example, the owner player may notice a delayed reaction after clicking the **Attack** button. This is because the message needs time to go around. Advanced multiplayer game processes, such as *prediction* and *lag compensation*, can be used to improve the player experience, but they are out of the scope of this book.

Based on the aforementioned changes to the newly added `APangaeaCharacter::Attack_Broadcast_RPC()` function, we can easily resolve the network synchronization issue for enemy attacks in `AEnemyController.cpp`. Enemies are only owned by the server, so all we need to do is to make enemies call `Attack_Broadcast_RPC()` when they decide to attack, as follows:

```
void AEnemyController::MakeAttackDecision(APawn* targetPawn)
{
    auto enemy = Cast<AEnemy>(GetPawn());
    auto dist = FVector::Dist2D(
targetPawn->GetActorLocation(),
GetPawn()->GetTargetLocation());

    if (dist <= enemy->AttackRange && enemy->CanAttack())
    {
        StopMovement();
        enemy->Attack_Broadcast_RPC();
    }
}
```

RPCs can be used to send messages and some parameter data across network connections, but they are not suitable for synchronizing actor variables when the variable values are changed. In this case, the replication mechanism is brought in to take care of synchronizing actor variables.

Syncing actor variables to clients with replications

Variable replication is a useful networking tool that you can use in Unreal to develop multiplayer games. `Actor` variables with a `Replicated` tag indicate that whenever their values are changed, they are automatically replicated (synced) to all connected remote proxies. For example, a character needs to replicate its `_HealthPoints` value from the server to all clients for displaying the health points bar.

We will create a `HealthBar` widget and add a replication notification for displaying the characters' health points after.

Open PangaeaCharacter.h and add the Replicated tag as the specifier of the UPROPERTY macro for the _HealthPoints variable, as follows:

```
UPROPERTY(Replicated)
int _HealthPoints;
```

We also want to declare the GetLifetimeReplicatedProps() overriding function. Unreal requires our code to explicitly return the replicated variables in this function. Here's how we can do this:

```
void GetLifetimeReplicatedProps(
    TArray<FLifetimeProperty>& OutLifetimeProps) const override;
```

In PangaeaCharacter.cpp, add the following code:

```
#include <Net/UnrealNetwork.h>
void APangaeaCharacter::GetLifetimeReplicatedProps(TArray<
FLifetimeProperty >& OutLifetimeProps) const
{
Super::GetLifetimeReplicatedProps(OutLifetimeProps);
DOREPLIFETIME(APangaeaCharacter, _HealthPoints);
}
APangaeaCharacter::APangaeaCharacter()
{
    PrimaryActorTick.bCanEverTick = true;
    bReplicates = true;
}
```

The preceding code snippet does the following three things:

- Includes the Net/UnrealNetwork.h header file to enable calling the DOREPLIFETIME macro

- The GetLifetimeReplicatedProps function's implementation calls the superclass's overridden function and the DOREPLIFETIME macro to replicate the _HealthPoints variable

- In the constructor of APangaeaCharacter, bReplicates is set to true to ensure the character and its networking variables will be replicated

Now, while playing the game, once a character's _HealthPoints variable is changed on the server, it will be replicated to update the corresponding character's _HealthPoints value on all connected clients.

The next thing we want to do is display characters' _HealthPoints value with a health bar.

Updating the character health bar with RepNotify

RepNotify indicates that a replicated variable can have a handler function, which is called when the variable's value changes. For example, if _HealthPoints is designated the RepNotify function OnHealthPointsChanged(), when the variable value is changed on the server and replicated to the clients, OnHealthPointsChanged() is then invoked on all connected clients.

So, we can update the character's health bar inside the RepNotify function.

> **Note**
>
> In comparison to RPCs, RepNotify is preferable because it helps simplify your code and uses less network bandwidth.

Creating the RepNotify handler function

To carry out corresponding processes on the client side when the value of a network-replicated property is updated from the server, we can associate a function with that property.

In *Pangaea*, to visualize health bar changes whenever any actor's _HealthPoints value is updated from the server, let's add OnHealthPointsChanged() as a UFUNCTION macro to the APangaeaCharacter class. So, in PangaeaCharacter.h, add this code:

```
UFUNCTION()
void OnHealthPointsChanged();
```

And in PangaeaCharacter.cpp, add this code:

```
void APangaeaCharacter::OnHealthPointsChanged()
{
//We will write code here to update the health bar
}
```

This OnHealthPointsChanged handler function should then be hooked up to the RepNotify event.

Hooking up the RepNotify handler function

We can use ReplicatedUsing to replace the old Replicated specifier. The new specifier allows _HealthPoints to designate OnHealthPointsChanged () as its RepNotify handler function, as follows:

```
UPROPERTY(Replicatedusing = OnHealthPointsChanged)
int _HealthPoints;
```

Now, we want to create a health bar for displaying character _HealthPoints values (remember that health bars are only required on the client side).

Creating the UHealthBarWidget class

To create a health bar for *Pangaea* characters, we first create a new UI widget class, UHealthBarWidget, which inherits from UUserWidget. This operation will generate new HealthBarWidget.h and HealthBarWidget.cpp files.

So, in the new header file, we want to define the UHealthBarWidget class, like so:

```
#pragma once
#include "CoreMinimal.h"
#include "Components/ProgressBar.h"
#include "Blueprint/UserWidget.h"
#include "HealthBarWidget.generated.h"

UCLASS()
class PANGAEA_API UHealthBarWidget : public UUserWidget
{
    GENERATED_BODY()
public:
UPROPERTY(VisibleAnywhere,
        BlueprintReadWrite,
        meta = (BindWidget))
    UProgressBar* HealthProgressBar;
};
```

The preceding code snippet defines the UHealthBarWidget class, which contains only one property: HealthProgressBar. The HealthProgressBar property stores the pointer of the associated **Progress Bar** UI component.

HealthProgressBar is a public property of type UProgressBar*, with the meta=(BindWidget) specifier. It implies that if a **Progress Bar** UI component is added to the widget blueprint with the same identifier (HealthProgressBar), it will automatically be bound to this property.

Since no function is declared in the header file, HealthBarWidget.cpp only needs one line of code, which simply includes the header file:

```
#include "HealthBarWidget.h"
```

Creating the BP_HealthBar blueprint

We now want to create a new UI widget, BP_HealthBar, which only has a progress bar on it.

Reopen the project in the Unreal Editor and create a BP_HealthBarWidget blueprint, saving it under **Content | Topdown | Blueprints**. Then, make the new blueprint a child of the UHealthBarWidget class, as seen in *Figure 10.8*:

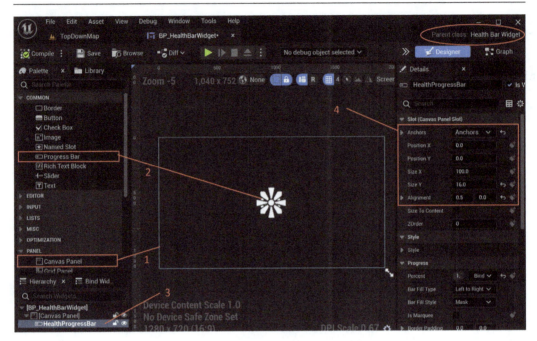

Figure 10.8 – Creating a BP_HealthBar.png blueprint

Let's explore the steps for creating the BP_HealthBar widget:

1.   Drag and drop **Canvas Panel** onto the scene.

2.   Drag and drop **Progress Bar** onto the scene.

3.   Change ProgressBar_0 to HealthProgressBar. This name has to be identical to the UHealthBarWidget::HealthProgressBar variable name.

4.   Select **HealthProgressBar** and set the attributes as shown in *Figure 10.8*.

Having followed the previous steps, we have created the BP_HealthBar widget. Now, we can add it and associate it with both the player and enemy characters.

Adding the HealthBar to BP_PlayerAvatar and BP_Enemy

Now, we need to add the BP_HealthBar UI widget to both BP_PlayerAvatar and BP_Enemy so that the health bar shows up above the character.

Open `BP_PlayerAvatar` and add a widget (see *Figure 10.9*):

Figure 10.9 – Adding a widget to BP_PlayerAvatar

Rename the added widget from **Widget** to a meaningful name – `HealthBar`, for example – and configure the attributes, like so:

- Set **Space** to **Screen** so that the health bar is always facing the player camera.

- Choose `BP_HealthBarWidget` for **Widget Class**.

- Make the **Draw Size** value **50** by **20**:

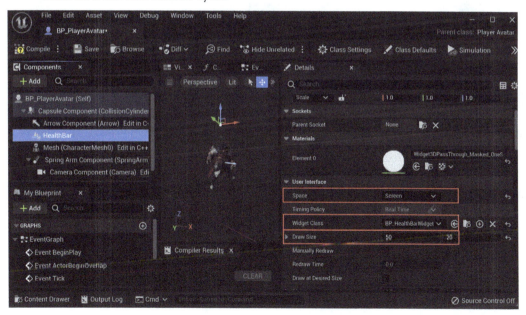

Figure 10.10 – Setting up the HealthBar widget for PlayerAvatar

You can follow the preceding steps to add `BP_HealthBarWidget` to `BP_Enemy`.

The next thing we want to do is to update the health bar when the character's health points value changes.

Updating the health bar when the RepNotify function is invoked

Now is the time to implement the `OnHealthPointsChanged()` function to update the progress bar on `BP_HealthBarWidget`. The idea is to add a `UUserWidget*`-type variable to the `APangaeaCharacter` class, and this variable should be both readable and writeable to blueprints.

Open `PangaeaCharacter.h` and add the following variable definition to the public section of the `APangaeaCharacter` class:

```
UPROPERTY(VisibleAnywhere, BlueprintReadWrite)
UUserWidget* HealthBarWidget;
```

Having defined the `HealthBarWidget` variable, we can get and assign the **Health Bar** widget to this variable in `BP_PlayerAvatar` and `BP_Enemy`, like so:

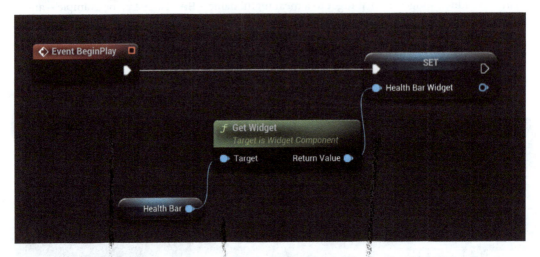

Figure 10.11 – Event Graph to set the Health Bar Widget reference for PlayerAvatar

The provided Event Graph achieves the assignment task through the following approach:

- When **BeginPlay** is triggered, the **Get Widget** node retrieves the **Health Bar** widget
- The **Set** node sets `APangaeaCharacter::HealthBarWidget` with the retrieved **Health Bar** widget instance

Let's write the implementation code for the OnHealthBarChanged() function in PangaeaCharacter.cpp, setting the progress bar value of HealthBarWidget whenever its value is updated from the server:

```cpp
void APangaeaCharacter::OnHealthPointsChanged()
{
  if (HealthBarWidget != nullptr)
  {
    float normalizedHealth = FMath::Clamp(
    (float)_HealthPoints / HealthPoints, 0.0f, 1.0f);
    auto healthBar = Cast<UHealthBarWidget>(HealthBarWidget);
    healthBar->HealthProgressBar->SetPercent(normalizedHealth);
  }

  if (_AnimInstance != nullptr)
  {
    _AnimInstance->State = ECharacterState::Hit;
  }

  if (IsKilled())
  {
    PrimaryActorTick.bCanEverTick = false;
  }
}
```

Let's break this code down:

- (float)_HealthPoints / HealthPoints calculates the current health percentage. It converts the current value of _HealthPoints into a float type and then divides it by the maximum HealthPoints value.

- The FMath::Clamp() math function clamps the float value between 0.0f and 1.0f.

- Since the type of HealthBarWidget is a UUserWidget* type, it has to be cast into a UHealthBarWidget* type so that its HealthProgressBar variable can be accessed.

- The HealthProgressBar->SetPercent() function is called to update the progress bar display.

- _AnimInsntance->State is set to be ECharacterState::Hit to play the **Hit** animation.

- PrimaryActorTick.bCanEverTick is set to false to stop the ticking functionality of this actor when it is killed.

Now, launch the game, and you should now see nice health bars appearing over the characters, as represented here:

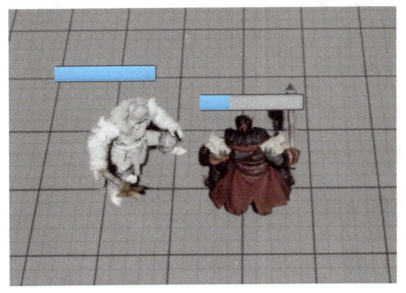

Figure 10.12 – Characters with health bars

One more thing we want to take care of is processing *hits* on the server because only the game server has the authority to change characters' `HealthPoints` values.

Processing hits on the server

In multiplayer games, only authoritative actors have control over their actor states. Unreal Engine by default uses a server-authoritative strategy for multiplayer games. The server is the place where the gameplay happens, and all the clients are just remote proxies that accept player inputs and reflect the current game to players.

Since the *Pangaea* server is an authoritative server, the hit process should only be handled on the server side. The changed implementation of the `Hit()` function of `APangaeaCharacter` is shown as follows:

```
void APangaeaCharacter::Hit(int damage)
{
    if (IsKilled())
    {
        return;
    }

    if (GetNetMode() == ListenServer::NM_Client
```

```
        && HasAuthority())
    {
        _HealthPoints -= damage;
        OnHealthPointsChanged();
    }
}
```

Let's take a look at the code:

- Calling AActor::GetNetMode() can get the game mode, which could be either NM_Client or NM_ListenServer in this case

- Calling AActor::HasAuthority() returns true when the actor is authoritative

- The code only reduces _HealthPoints when it is running on the server and the actor is authoritative

- Calling OnHealthPointsChanged() on the listen server processes the RepNotify function locally

The last issue we want to resolve in this chapter is the non-visibility and non-movement of fireballs on the client side.

Spawning fireballs on the server side

Since the multiplayer *Pangaea* game is built on the server-authoritative approach, fireballs should be spawned and pooled on the server side rather than on the client side. It is very easy to change the code in ADefenseTower::Tick() to avoid spawning fireballs on the client side, as seen here:

```
void ADefenseTower::Tick(float DeltaTime)
{
    Super::Tick(DeltaTime);

    if (_Target != nullptr &&  GetNetMode() != NM_Client)
    {
        Fire();
    }
}
```

To show and move active fireballs on clients, you can open `BP_FireBall` and make sure that the **Replicates** and **Replicate Movement** boxes are checked:

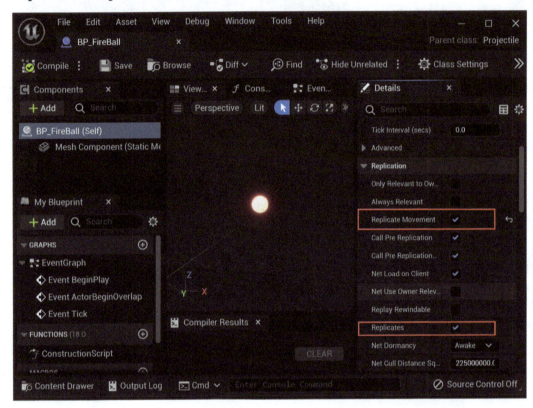

Figure 10.13 – Enabling fireball replication

In the editor, launch and enjoy playing the multiplayer version of the *Pangaea* game:

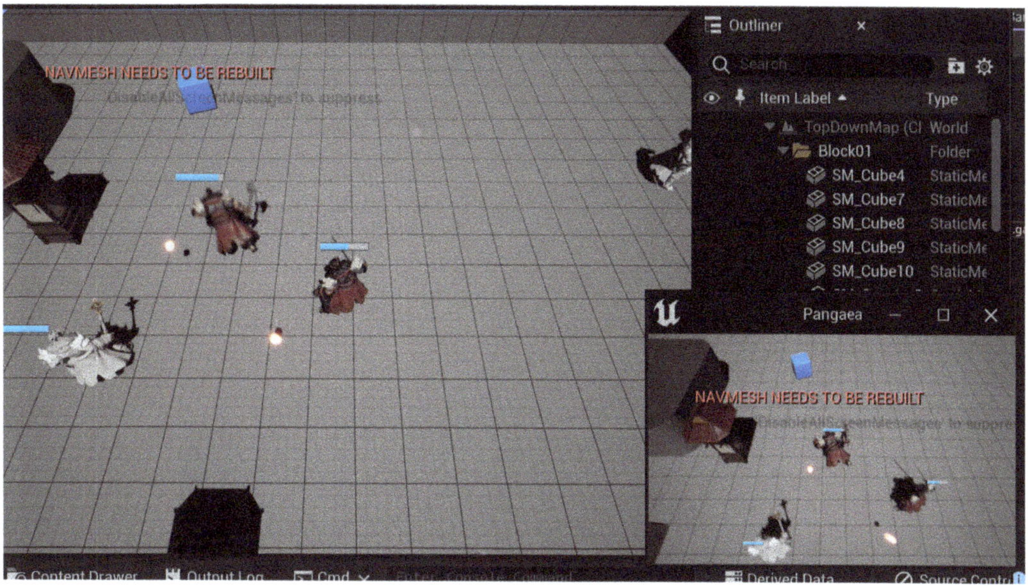

Figure 10.14 – Playing the multiplayer Pangaea game

At this point, you should have acquired essential concepts and techniques for creating an immersive and interactive multiplayer game in Unreal.

Summary

This chapter demonstrated how to make a single-player game a multiplayer game in Unreal. The content we introduced in this chapter focuses on learning essential concepts and methods to follow to solve common multiplayer game issues. The knowledge you gained from this chapter can be a starting point that will help you continue learning about advanced multiplayer game development in the future.

In the first section, we introduced the concepts of single-player, local multiple, and network multiplayer games so that you got a basic sense of what is required for multiplayer gameplay.

Then, you learned how to launch the *Pangaea* game with two players (one listen server and one client) in the editor. Based on that, we explained multiplayer game net modes as well as identified some issues associated with multiplayer gameplay.

To solve issues in the multiplayer version game, you learned to use RPCs to notify the server and the other clients of attack actions, replicate actor states (_HealthPoints, for example) to clients, handle the RepNotify variable to update the character health bar, and eventually spawn fireballs and process hits on the server.

At the end of this chapter, we have a playable multiplayer game.

In the next chapter, we are going to develop some UI screens to control the game flow that allows players to host a game server or connect as a client to join the game session, disconnect to leave the session, and exit the game.

11

Controlling the Game Flow

In the previous chapter, we completed the core gameplay for *Pangaea*. To transform it into a fully fledged game, we will incorporate a basic game flow that enables players to go into the game through the main menu and return to the main menu upon exiting the game.

A game's flow control could be very complex, depending on the game's design. Many games use a centralized control system, such as a **finite-state machine** (**FSM**), to control transitions from one game state to another, but using the advanced control system is outside the scope of this book.

To make it easier to learn the C++ scripting skills to control the game's flow, we will design and implement a minimum game flow for the *Pangaea* game. We will do this by covering the following topics:

- Designing the *Pangaea* game's flow
- Creating the UI widgets
- Adding networking functions to `PangeaGameInstance`
- Adding UI widgets to game levels
- Adding the game timer
- Destroying a base defense tower to win the game

Technical requirements

The code for this chapter can be found at `https://github.com/PacktPublishing/Unreal-Engine-5-Game-Development-with-C-Scripting/tree/main/Chapter11`.

Designing the Pangaea game's flow

To minimize the complexity of the *Pangaea* game's flow, the game is designed to have only one lobby (*LobbyMap*) and one gameplay (*TopDownMap*) level. The lobby level shows the main menu, which allows players to choose to play as a game host or a client, whereas the gameplay level is the map on which players fight. Here is the game flow chart:

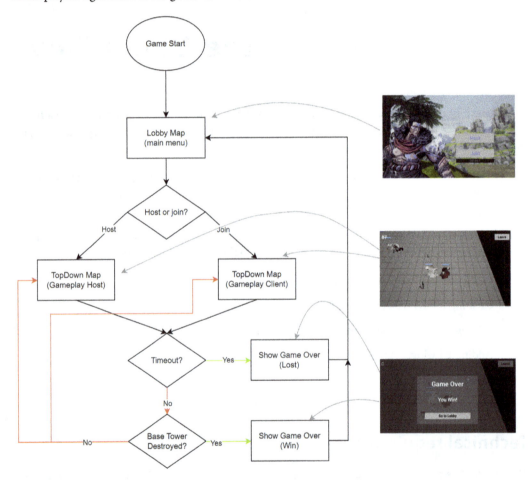

Figure 11.1 – Pangaea game flow diagram

Based on this flow chart, the next thing we need to do is create the three UI widgets for displaying the main menu, the HUD, and the **Game Over** window, respectively.

Creating the UI widgets

To create UI widgets, first, select the **All | Content | TopDown | Blueprints** folder (or wherever you want to place the new widgets in the **Content Drawer**). Then, right-click in the empty area and choose **User Interface | Widget Blueprint** from the pop-up menu:

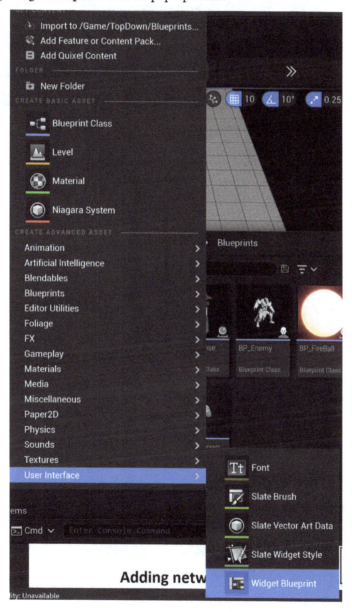

Figure 11.2 – Creating a UI widget from the editor pop-up menu

Now, click the **User Widget** button to pick the root widget:

Figure 11.3 – Clicking the User Widget button to create the widget

After picking the root widget, you should see a new item named **NewWidgetBlueprint** in the folder:

Figure 11.4 – The new widget blueprint

Rename the new widget to the name you want – I will choose `BP_LobbyWidget`. Then, repeat these steps and create two more widgets called **BP_HUDWidget** and **BP_GameOver**.

Now, let's edit these widgets' settings.

Creating BP_LobbyWidget

With **BP_LobbyWidget** created, double-click to open it in the Widget Editor. Now, you can design the user interface, as seen in *Figure 11.5*. To do this, follow these steps:

1. Drag and drop a **Canvas Panel** property onto the screen design area.

2. Place an **Image** control onto the screen with the following settings:

    - **Name:** `BGImage`

    - **Anchors:** `Stretch Both`

- **Alignment**: $(0.0, 0.0)$

- **Brush | Image**: `LobbyBG` (you can download `LobbyBG.png` from this book's `GitHub/PangaeaAssets` folder and import it into the project)

3. Add a button to the screen with the following settings:

    - **Name**: `ButtonHost`

    - **Alignment**: $(0.0, 0.0)$

    - **Anchors**: `Center`

    - **Position X**: `1000.0`

    - **Position Y**: `500.0`

    - **Size X**: `500.0`

    - **Size Y**: `120.0`

4. Add a **TextBox** control as the child of **ButtonHost** with the following settings:

    - **Horizontal Alignment**: `Center Align Horizontally`

    - **Vertical Alignment**: `Center Align Vertically`

    - **Text**: `"Host"`

    - **Color and Opacity** (RGBA): $(0.04, 0.15, 0.5, 1.0)$

    - **Font**:

        - **Font Family**: `Roboto`

        - **Type Face**: `Bold`

        - **Size**: `48`

5. Add one more button to the screen with the following settings:

    - **Name**: `ButtonJoin`

    - **Anchors**: `Center`

    - **Alignment**: $(0.0, 0.0)$

    - **Position X**: `1000.0`

    - **Position Y**: `700.0`

    - **Size X**: `500.0`

    - **Size Y**: `120.0`

6. Add a **TextBox** control as the child of **ButtonJoin** and set it up with the same settings as *step 4*, except substitute **Host** with **Join**.

7. Add an **EditableTextBox** control to the screen with the following settings:

   - **Name**: InputAddress

   - **Anchors**: Center

   - **Alignment**: (0.0, 0.0)

   - **Position X**: 1000.0

   - **Position Y**: 700.0

   - **Size X**: 500.0

   - **Size Y**: 120.0

   - **Text**: "127.0.0.1" (the default is the local server IP address)

   - **Justfication**: Align Text Center

Again, you can see the result in *Figure 11.5*:

Figure 11.5 – Designing BP_LobbyWidget in the Widget Editor

Creating BP_HUDWidget

Now, with **BP_HUDWidget** created, open it in the Widget Editor. Then, design the user interface as seen in *Figure 11.6*. To do so, follow these steps:

1. Drag and drop a **Canvas Panel** property onto the screen design area.

2. Place a **TextBox** control in the top-left corner with the following settings:

   - **Name**: Timer
   - **Anchors**: Top-left
   - **Alignment**: (0.0, 0.0)
   - **Position X**: 50.0
   - **Position Y**: 50.0
   - **Size X**: 300.0
   - **Size Y**: 60.0
   - **Justification**: Align Text Left
   - **Color and Opacity** (RGBA): (0.04, 0.15, 0.5, 1.0)
   - **Font**:
     - **Font Family**: Roboto
     - **Type Face**: Bold
     - **Size**: 48

3. Add a button to the screen with the following settings:

   - **Name**: ButtonLeave
   - **Anchors**: Top-right
   - **Alignment**: (1.0, 0.0)
   - **Position X**: -50.0
   - **Position Y**: 50.0
   - **Size X**: 200.0
   - **Size Y**: 80.0

4. Add a **TextBox** control as the child of **ButtonLeave** with the following settings:

- **Horizontal Alignment**: Center Align Horizontally
- **Vertical Alignment**: Center Align Vertically
- **Text**: "Leave"
- **Color and Opacity** (RGBA): (0.04, 0.15, 0.5, 1.0)
- **Font**:

  - **Font Family**: Roboto
  - **Type Face**: Bold
  - **Size**: 48

Again, you can see the result in *Figure 11.6*:

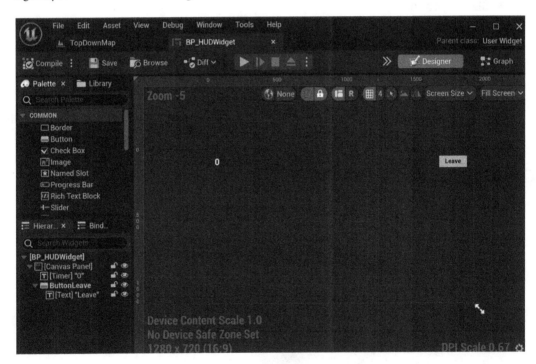

Figure 11.6 – Designing BP_HUDWidget in the Widget Editor

Creating BP_GameOverWidget

For the last widget, with **BP_GameOverWidget** created, open it in the Widget Editor. Then, design the user interface as seen in *Figure 11.7*. To do so, follow these steps:

1. Drag and drop a **Canvas Panel** property onto the screen design area.

2. Place an **Image** control on the screen with the following settings:

    *   **Name**: Background

    *   **Anchors**: Stretch Both

    *   **Alignment**: (0.0, 0.0)

    *   **Brush | Image**: None

    *   **Color and Opacity** (RGBA): (0.0, 0.0, 0.0, 0.5)

3. Drag and drop another **Image** control on the screen with the following settings:

    *   **Name**: Panel

    *   **Anchors**: Center

    *   **Position X**: 0.0

    *   **Position Y**: 0.0

    *   **Alignment**: (0.5, 0.5)

    *   **Brush | Image**: None

    *   **Color and Opacity** (RGBA): (0.2, 0.2, 0.2, 0.5)

4. Add a **TextBox** control to the screen with the following settings:

    *   **Name**: Title

    *   **Anchors**: Center

    *   **Alignment**: (0.5, 0.5)

    *   **Position X**: 0.0

    *   **Position Y**: -200.0

    *   **Size X**: 800.0

    *   **Size Y**: 100.0

    *   **Justification**: Align Text Left

    *   **Color and Opacity** (RGBA): (1.0, 1.0, 1.0, 1.0)

- **Font:**

  - **Font Family:** Roboto
  - **Type Face:** Bold
  - **Size:** 64

5. Add one more **TextBox** control to the screen with the following settings:

- **Name:** Result
- **Anchors:** Center
- **Alignment:** (0.5, 0.5)
- **Position X:** 0.0
- **Position Y:** 0.0
- **Size X:** 800.0
- **Size Y:** 80.0
- **Justification:** Align Text Left
- **Color and Opacity** (RGBA): (1.0, 1.0, 1.0, 1.0)
- **Font:**

  - **Font Family:** Roboto
  - **Type Face:** Bold
  - **Size:** 48

6. Add a button to the screen with the following settings:

- **Name:** ButtonLobby
- **Anchors:** Center
- **Alignment:** (0.5, 0.5)
- **Position X:** 0.0
- **Position Y:** 200.0
- **Size X:** 600.0
- **Size Y:** 80.0

7. Add a **TextBox** control as the child of **ButtonLobby** with the following settings:

-   **Horizontal Alignment**: `Center Align Horizontally`

-   **Vertical Alignment**: `Center Align Vertically`

-   **Text**: `"Go to Lobby"`

-   **Color and Opacity** (RGBA): `(0.0, 0.0, 0.0, 1.0)`

-   **Font**:

    -   **Font Family**: `Roboto`

    -   **Type Face**: `Bold`

    -   **Size**: `32`

The results can be seen in *Figure 11.7*:

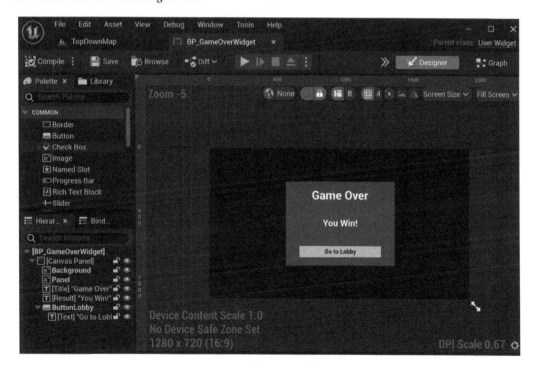

Figure 11.7 – Designing BP_GameOverWidget in the Widget Editor

While designing the three UI widgets, we added four buttons to the screen: **ButtonHost**, **ButtonJoin**, **ButtonLeave**, and **ButtonLobby**. To make these four buttons functional, we need to create three functions (`StartListenServer`, `JoinAsClient`, and `LeaveGame`) and hook them to these button events (**ButtonLeave** and **ButtonLobby** share the `LeaveGame` function).

Adding networking functions to PangaeaGameInstance

Before we add these functions to `APangaeaGameInstance`, it is important to explain why we must add these functions to this class but somewhere else.

We know that `GameInstance` only exists on the client side and that the **ButtonHost**, **ButtonJoin**, **ButtonLeave**, and **ButtonLobby** buttons are pressed by players when they are playing the game on their end, so it makes sense to implement those functions that allow players to host, join, and leave games on the client side.

Now, let's open the `PangaeaGameInstance.h` file and add the following code to declare the functions in the `ApangaeaGameInstance` class:

```
public:
UFUNCTION(BlueprintCallable, Category = "Pangaea")
  void StartListenServer();
  UFUNCTION(BlueprintCallable, Category = "Pangaea")
  void JoinAsClient(FString IPAddress);
  UFUNCTION(BlueprintCallable, Category = "Pangaea")
  void LeaveGame();
```

All the functions are marked with the `BlueprintCallable` specifier, which indicates that these functions can be called from blueprints.

Now, we can implement these functions in `PangaeaGameInstance.cpp`:

```
#include "PangaeaGameInstance.h"
#include "Kismet/GameplayStatics.h"

void UPangaeaGameInstance::StartListenServer()
{
auto world = GEngine->GetCurrentPlayWorld();
  UGameplayStatics::OpenLevel(world,
        "TopDownMap", true, "?listen");
}
```

```
void UPangaeaGameInstance::JoinAsClient(FString IPAddress)
{
  auto world = GEngine->GetCurrentPlayWorld();
  UGameplayStatics::OpenLevel(world,
                  *IPAddress, true, "?join");
}
void UPangaeaGameInstance::LeaveGame()
{
  auto world = GEngine->GetCurrentPlayWorld();
  UGameplayStatics::OpenLevel(world, "LobbyMap");
}
```

Let's understand this code a little more:

- All three functions call GEngine->GetCurrentPlayWorld() at the beginning.

- GEngine is a global engine pointer that allows developers to access the engine's core information anywhere at runtime. GetCurrentPlayWorld, GetFirstLocalPlayerController, GetGamePlayers, and so on are the most useful member functions.

- UGameplayStatics is a useful static engine class that provides utility functions that can be called from both C++ and Blueprint. OpenLevel, GetGameMode, GetGameState, and GetGameInstance are frequently called functions.

- UGameplayStatics::OpenLevel(world, "TopDownMap", true, "?listen"); starts the game as a listen server and travels to **TopDown**, which is the gameplay level.

- UGameplayStatics::OpenLevel(world, *IPAddress, true, "?join"); starts the game as a client and connects to the server. The server's IP address is provided by the second parameter, IPAddress.

To convert an FString's IPAddress value into an FName variable, add * before the variable's name (*IPAddress). The OpenLevel function's second parameter, IPAddress, is of the FName type, while the JoinAsClient caller function is an FString type with a value of IPAddress.

Now, we can open the UI widgets we just created and hook up these functions to the corresponding events. For example, the **Host** button on **BP_LobbyWidget** should trigger the OnClick event, and then call the StartListenServer() function. The **Join** button retrieves a string value from the **InputIPAddress** text box and uses it as a parameter when calling the JoinAsClient() function:

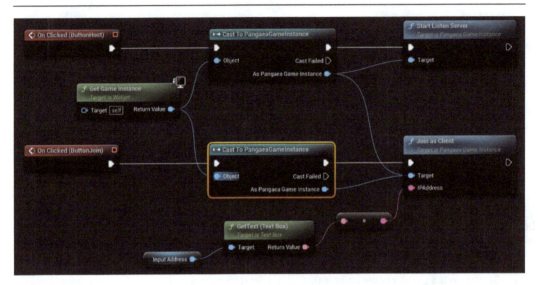

Figure 11.8 – Handling the Host and Join buttons' OnClick events on BP_LobbyWidget

Let's break this down a bit more:

- The **On Clicked** nodes are the events that are triggered when the **Host** or **Join** button is pressed.

- Once one of the two events is trigged, the **Cast To PangaeaGameInstance** node gets the current system's game instance and casts it so that it's a `PangaeaGameInstance` type of instance. This ensures that the graph can access and call its member functions

- The **Start Listen Server** node calls the `ApangaeaGameInstance` class's member C++ function, `StartListenServer`.

- The **Join As Client** node calls the `ApangaeaGameInstance` class's C++ member function, `JoinAsClient`.

- The **Get Text** node fetches the text from the **Input Address** input box. Then, the text is converted into an `FString` value, which is then passed as a parameter to the **Join As Client** node.

Follow a similar process to make both the **Leave Game** and **Go to Lobby** button events call the `ApangaeaGameInstance` class's `LeaveGame` member function when the events are triggered.

Open `BP_HUDWidget` and add the following blueprint graph to handle the **On Clicked (ButtonLeave)** event:

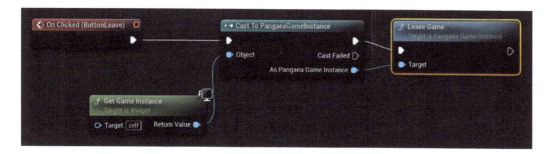

Figure 11.9 – Handling the LeaveGame button event to navigate to the lobby

Then, open BP_GameOverWidget and add the following blueprint graph to handle the **On Clicked (ButtonLobby)** event:

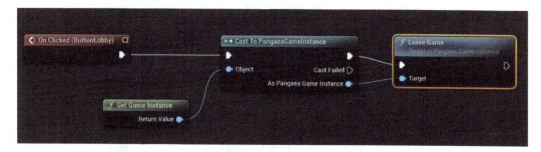

Figure 11.10 – Handling the GoToLobby button event to navigate to the lobby

With the multiplayer UI operations connected, our next objective is to change **Multiplayer Options** in the **Player Mode and Player Settings** menu. The **Number of Players** value can remain set to **2**, but the **Net Mode** can now be changed to **Play Standalone**:

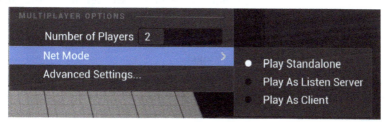

Figure 11.11 – Changing the multiplayer options

Since we've already written code and started the game as either a listen server or a client from the lobby, we no longer rely on the engine to automatically launch the multiplayer mode game windows and establish the connection. By changing **Net Mode** to **Play Standalone**, we can launch two standalone games when clicking the **Play** button.

Now that we've created three UI widgets called **BP_LobbyWidget**, **BP_HUDWidget**, and **BP_GameOverWidget**, we can incorporate them into the corresponding game levels.

Adding UI widgets to game levels

UI widgets can be created and added to the viewport of the current game level, but the preferred place to create UI widgets is in Level Blueprints. So, click on the **List of World Blueprints** button on the toolbar and choose **Open Level Blueprint** to open and edit the Level Blueprint:

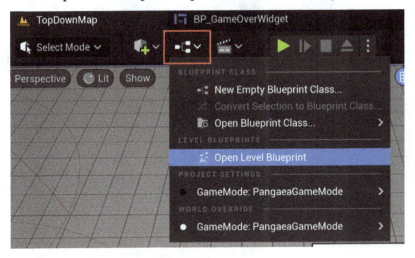

Figure 11.12 – Opening the current Level Blueprint

Next, open **LobbyMap** and edit the Level Blueprint as follows:

1. Search for and add a **Create Widget** node to the graph. The new node's title only displays **Construct NONE** at the moment.

2. Select **BP_LobbyWidget** from the **Class** drop-down menu. The **Create Widget** node's title will now display **Create BP Lobby Widget Widget**.

3. Add a **Add to View Port** node and connect the output pin of the **Create BP Lobby Widget Widget** node to the input pin of **Add to View Port** node. Connect the execution pins of these two nodes as well:

Figure 11.13 – Creating and showing BP_LobbyWidget on LobbyMap

Next, open **TopDownMap** and edit the Level Blueprint the same way we just did for **LobbyMap**. Here, select **BP_HUDWidget** from the drop-down list under **Class**:

Figure 11.14 – Creating and showing BP_HUDWidget on TopDownMap

Now, if you want to play the game, remember to open **LobbyMap** as the current level. When you launch the game in the editor, you should see two game instances – one in the editor's viewport and another in a separate window:

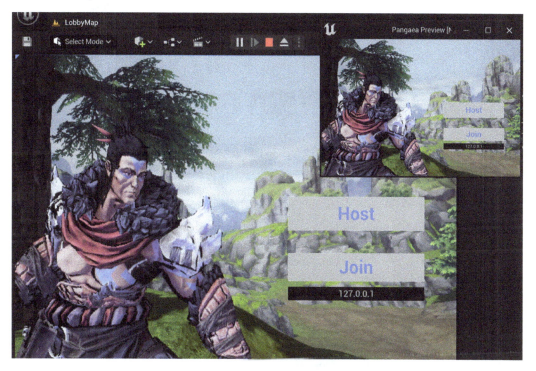

Figure 11.15 – Launching two game instances

On one game screen, click the **Host** button to start the game as a listen server. Then, click the **Join** button in the other window to join the game. We currently don't need to change the server IP address because we are running the server and the client on the same machine, which means that it is a local server (the IP address is 127.0.0.1).

While playing the game, the game HUD will display the Timer property in the top-left corner and the **Leave** button in the top-right corner:

Figure 11.16 – Playing the game with the HUD screen displayed

The next thing we will do is process how the game ends. The idea is that when the game's timer counts down to 0, the players lose the game; otherwise, if the designated base tower is destroyed, the players win the game. So, let's add a game timer and its gameplay mechanism to the game so that it can control when to end the game.

Adding the game timer

The game timer serves the purpose of counting down and restricting the duration of a game session. To make the timer work, follow these steps:

- Introduce a new float variable named Timer into the APangaeaGameState class.

- Ensure that the Timer variable is capable of being replicated to inform clients about changes to the timer's value.

- Define a C++ Delegate that can be linked to Blueprint events, enabling the execution of an event function whenever the timer's value changes. This function will update the display for the countdown.

- Create a custom event on **BP_HUDWidget** and bind the customer event to the C++ Delegate. In the meantime, link the custom event to an event function, which will be responsible for updating the display.

- Count down the Timer variable within the Level Blueprint of **TopDownMap**.

- Display the **Game Over** window once the timer reaches 0.

Let's get started!

Adding the Timer variable to the APangaeaGameState class

The rationale behind including the Timer variable in APangayaGameState is its presence on both the server and client sides – any game-related states that are shared between all connected clients should be properties of this class.

Ensure that you have already created the APangaeaGameState class, which is derived from the AGameState class. Then, open the PangaeaGameState.h file and insert the following code snippet:

```
public:
void GetLifetimeReplicatedProps(
  TArray<FLifetimeProperty>& OutLifetimeProps) const override;

UPROPERTY(BlueprintReadWrite, Category = "Pangaea")
float Timer = 0;
```

Here, the GetLifetimeReplicatedProps function is needed to replicate variables, and the BlueprintReadWrite specifier indicates that blueprints can read and write this variable.

Next, add the implementation of the GetLifetimeReplicatedProps function to Pangaea GameState.cpp:

```
void APangaeaGameState::GetLifetimeReplicatedProps(TArray<
FLifetimeProperty >& OutLifetimeProps) const
{
    Super::GetLifetimeReplicatedProps(OutLifetimeProps);
    DOREPLIFETIME(APangaeaGameState, Timer);
}
```

The DOREPLIFETIME macro generates the necessary code to synchronize the Timer value across the network for all the clients.

Making the Timer variable replicable

To make the `Timer` variable replicable, we can add the `ReplicatedUsing` specifier and the `OnTimerChanged` notification handler function, like so:

```
public:
  UFUNCTION(BlueprintCallable, Category = "Pangaea")
  void OnTimerChanged();

  UPROPERTY(BlueprintReadWrite, ReplicatedUsing = OnTimerChanged,
    Category = "Pangaea")
  float Timer = 0;
```

`Timer` counts down only on the server side and will be synced to the clients. When the value of `Timer` is changed and replicated to the clients, the `OnTimerChanged` function will be called.

`OnTimerChanged` is a `BlueprintCallable` function because it needs to be called from the Level Blueprint on the listen server. Replication notifications are not fired on the server side, so to update the timer's display on the host, the notification's handler function should be directly called on the listen server once the timer's value is changed.

Defining OnTimeChangedDelegate

We already know how to use the `BlueprintCallable` specifier to make a C++ function callable to Blueprints. On the contrary, a C++ `Delegate` can be bound with one or more Blueprint custom events, which makes it possible to call custom Blueprint event functions from C++.

We can use the C++ DECLARE_DYNAMIC_MULTICAST_DELEGATE_OneParam macro to define the `FOnTimerChangedDelegate` delegate, and then use this delegate to create the `Delegate` variable, like so:

```
DECLARE_DYNAMIC_MULTICAST_DELEGATE_OneParam
  (FOnTimerChangedDelegate, float, Timer);

UCLASS()
class PANGAEA_API APangaeaGameState : public AGameStateBase
{
  public:
  UPROPERTY(BlueprintAssignable, Category = "Pangaea")
  FOnTimerChangedDelegate OnTimerChangedDelegate;

  …
}
```

So, the DECLARE_DYNAMIC_MULTICAST_DELEGATE_OneParam macro defines the FOnTimerChangedDelegate delegate (a Delegate name must start with an uppercase "F"), and the Delegate function should have only one float type parameter.

Plus, the BlueprintAssignable specifier indicates that Blueprint custom events can be bound to the OnTimerChangedDelegate variable.

Creating and binding the custom event to OnTimeChangedDelegate

Now, open **BP_HUDWidget** and edit the Event Graph to complete the following three tasks:

1. Get the OnTimerChangedDelegate node and bind it to the custom event, OnTimerCchangedEvent.
2. Update and display the timer.
3. Show the **Game Over** screen and pause the game.

You can see the results in *Figure 11.17*:

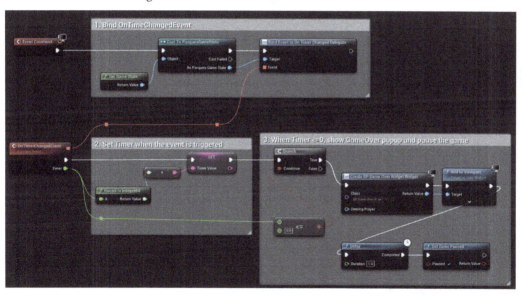

Figure 11.17 – Binding a custom event to OnTimerChangedDelegate on BP_HUDWidget

Let's investigate the details of each part of the graph.

Binding OnTimerChangedEvent to OnTimerChangedDelegate

The first part of the Event Graph gets `GameState` and casts it to `PangaeaGameState` first. Then, it uses the **Bind Event** node to bind to `OnTimerChangedDelegate`:

Figure 11.18 – Binding OnTimerChangedEvent to OnTimerChangedDelegate

Setting and displaying Timer

In the second part of the Event Graph, the parameter of `OnTimerChangedEvent` is the new `Timer` value, which is rounded to an integer number (we don't want to display the decimal part of the value on screen), and then converted into a text value that acts as the input of the **Set Timer** node:

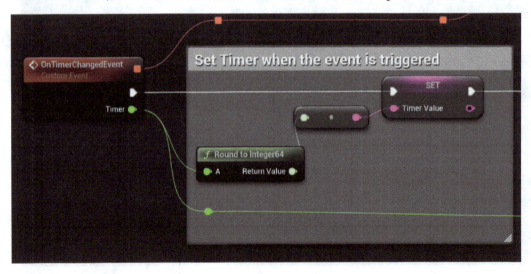

Figure 11.19 – Setting and displaying Timer

Opening the Game Over window and pausing the game

The third part of the Event Graph checks whether the `Timer` value is smaller than or equal to 0; if the result is true, it means that time has run out and the game is over:

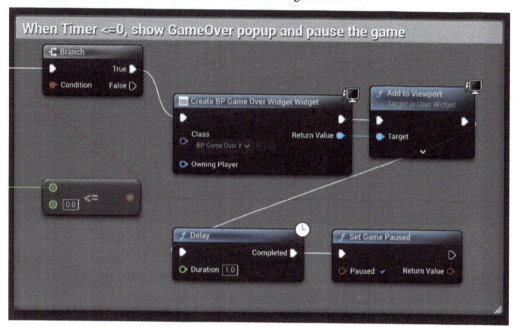

Figure 11.20 – Showing the GameOver window and pausing the game

Once the game is over, the **Create Widget** node creates **BP_GameOverWidget** and adds it to the viewport. Unreal allows you to stack multiple layers of UI widgets on the current viewport; in this case, **BP_GameOverWidget** is stacked in front of **BP_HUDWidget**.

The **Delay** node delays pausing the game for 1 second to allow the game to update displays and avoid rapidly pausing the game.

The **Set Game Paused** node's **Paused** box is checked, so it pauses the game.

With all that done, let's count down the timer while the game is running.

Counting down the timer

To implement the timer's counting down job, we can edit the **TopDownMap** Level Blueprint. The initialization of **Timer** is linked to the **BeginPlay** event:

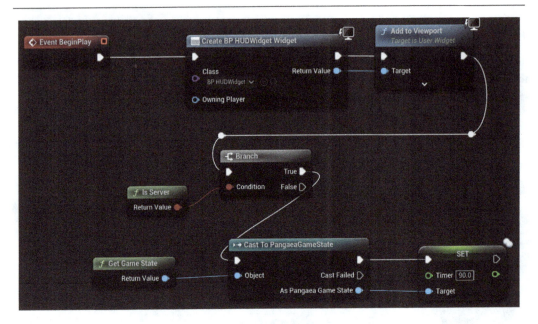

Figure 11.21 – Setting the Timer node's initial value

Here, the **Is Server** node returns `true` when the game is running on the server. The reason why we check it here is that we only want to set the initial value of **Timer** on the server.

Once we've confirmed that the game is running as a server, the **Set** node sets the **PangaeaGameState** node's `Timer` variable to 90 seconds.

Now, to count down the **Timer**, we need to create the following graph and establish its connection to the **Tick** event node:

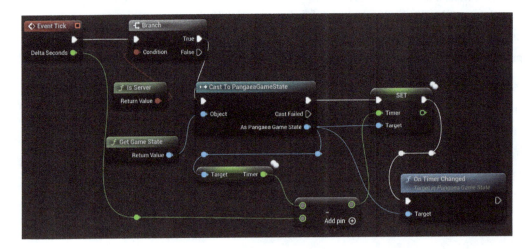

Figure 11.22 – Counting down the Timer variable

The graph first checks whether the game is running as a server.

If it is a server, the `Timer` variable's current value is retrieved from **PangaeaGameState**; then the **Subtraction** node subtracts **Delta Seconds** from the `Timer` value and sets the result back to **Timer**.

Once the new value is set to **Timer**, **OnTimerChanged** is called to notify the game (remember, **OnTimerChanged** is called by the timer replication notification on clients).

With that, the **Timer** node is working, but it is updated every frame, which unnecessarily consumes CPU time and network bandwidth too frequently. Let's learn more about this issue and try to improve it.

Since the **Tick** event is triggered every frame by default, this means this event function is called every frame. If the frame rate is 60, then the game updates **Timer** every 1/60 second. This seems to be unnecessarily too frequent, which may impact the game's performance and networking bandwidth because it is enough to be updated and synced for all the players once per second.

To improve this, we can simply open the **TopDownMap** node's Level Editor and change **Tick Interval (secs)** from **0.0** to **1.0**:

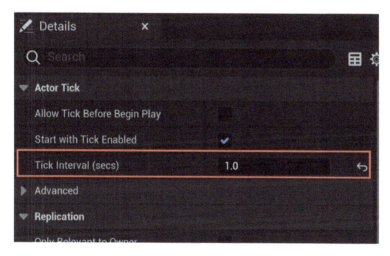

Figure 11.23 – Setting the TopDownMap node's Level Blueprint Tick Interval (secs) to 1 second

Designating APangaeaGameState as the project's game state class

One more thing we need to do is designate the game project's game state class – **ApangaGameState** should be used to substitute the original **AGameState** class.

To complete this task, we can simply add one more line of code at the end of the constructor of the `APangayaGameMode` class:

```
APangaeaGameMode::APangaeaGameMode()
{
    ...

    GameStateClass = APangaeaGameState::StaticClass();
}
```

Since the game is designed as an online co-op game, and the game has a timer to count down, once the timer reaches 0, the players lose the game. But how do players win the game? The idea is to designate one defense tower as the base, and once the base is destroyed, the players win.

Sound like a good idea? Let's make some changes to the `DefenseTower` class and get things up and running.

Destroying a base defense tower to win the game

To complete the implementation as well as add the `IsBase` tower's designation support for the `ADefenseTower` class, we must do the following:

- Add the `IsBase` flag to `ADefenseTower`.
- Modify the `Hit` function so that it deals with the server and client processes when the tower is destroyed.
- Add the `GameWin` flag and the `OnGameWin` function to `APangaeaGameState`.
- Add `OnGameWinLoseDelegate` to `APangaeaGameState`.
- Edit the `BP_DefenseTower` blueprint to bind `OnGameWinLoseEvent` to `OnGameWinLoseDelegate`.
- Show win or lose information in the **Game Over** window.

So, let's get started. First, open the `DefenseTower.h` file and add the `IsBase` flag to `AdefenseTower`. Then, add the following code:

```
public:
    ...

    UPROPERTY(EditAnywhere, Category = "Tower Params")
    bool IsBase = false;
```

Next, in DefenseTower.cpp, modify the Hit function so that it deals with the server and client processes when the tower is destroyed, like so:

```cpp
void ADefenseTower::Hit(int damage)
{
  if (IsKilled())
  {
    return;
  }

  if(HasAuthority())
  {
    _HealthPoints -= damage;
    OnHealthPointsChanged();

    if (_HealthPoints <= 0)
    {
      if (IsBase)
      {
        APangaeaGameState* gameState = Cast<APangaeaGameState>
          (UGameplayStatics::GetGameState(GetWorld()));
        gameState->OnGameWin();
      }
      else
      {
        Destroy();
      }
    }
  }
}

bool ADefenseTower::IsKilled()
{
  return (HealthPoints <= 0.0f);
}
```

The Hit function only reduces _HealthPoints. When _HealthPoints reaches 0, it is the server's responsibility to change the GameWin flag to true and call the OnGameWin function and process it on the server side. This is because only the server possesses the necessary authority.

Once a base tower is destroyed (_HealthPoints <= 0), the Hit function calls the APangaeaGameState::OnGameWin function, and the OnGameWin function sets the GameWin flag and notifies all the clients.

Once a regular tower is destroyed, the `Hit` function simply removes the tower by calling the `Destroy` function.

Now, we must add the `GameWin` flag and the `OnGameWin` function to `APangaeaGameState` in the `PangaeaGameState.h` file:

```
Public:
...

UPROPERTY(BlueprintReadWrite, ReplicatedUsing = OnGameWin,
  Category = "Pangaea")
bool GameWin;

UFUNCTION(BlueprintCallable, Category = "Pangaea")
void OnGameWin();
```

Let's break this down quickly:

- The `GameWin` variable is tagged so that it can be accessed by Blueprints
- `GameWin` is also tagged as a replicable variable, and the value change notification is hooked up with the `OnGameWin` function
- `OnGameWin` is tagged as a `BlueprintCallable` function, which means that Blueprints can call this function

The following is the `OnGameWin` function's implementation in the `PangaeaGameState.cpp` file:

```
void APangaeaGameState::OnGameWin()
{
    GameWin = true;
    OnGameWinLoseDelegate.Broadcast(true);
}
```

The next step is to add `FOnGameWinLoseDelegate` to `ApangaeaGameState`. This delegate type is used to define the `OnGameWinLoseDelegate` delegate variable, which is an interface that allows C++ to call blueprint functions:

```
DECLARE_DYNAMIC_MULTICAST_DELEGATE_OneParam(
  FOnGameWinLoseDelegate, bool, Win);

UCLASS()
class PANGAEA_API APangaeaGameState : public AGameStateBase
{
public:
    ...
```

```
    UPROPERTY(BlueprintAssignable, Category = "Pangaea")
    FOnGameWinLoseDelegate OnGameWinLoseDelegate;
}
```

The previous code uses the same method that we used to create `OnTimerChangedDelegate`.

We can now edit `BP_DefenseTower` to bind the `OnGameWinLoseEvent` event function to C++'s `OnGameWinLoseDelegate`. Open **BP_DefenseTower** to create the graph, as shown in *Figure 11.21*:

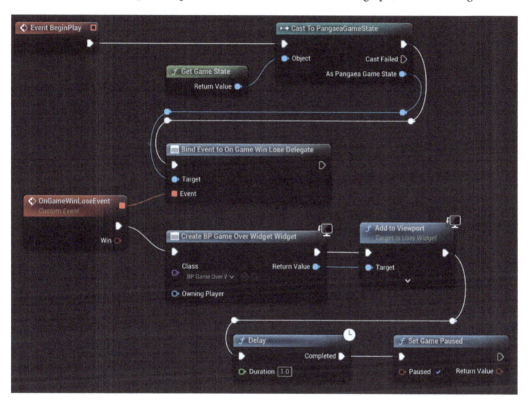

Figure 11.24 – Binding OnGameWinLoseEvent to OnGameWinLoseDelegate

This graph binds `OnGameWinLoseEvent` to `OnGameWinLoseDelegate` when the **BeginPlay** event occurs. Once `OnGameWinLoseEvent` is triggered, **BP_GameOverWidget** is created and added to the viewport to be displayed. The game then delays for 1 second and is paused.

Since **BP_DefenseTower** still doesn't have its health bar, we can add a **Widget** component to the **Components** hierarchy and name it `HealthBar`:

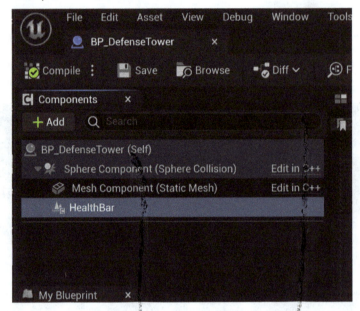

Figure 11.25 – Adding the tower's HealthBar

Then, fill in the following settings in the widget component's **Details** window (see *Figure 11.26*):

- **Variable Name:** `HealthBar`

- **Location:** `(0.0,  0.0, 380.0)`

- **Space:** `Screen`

- **Widget Class**: `BP_HealthBarWidget`:

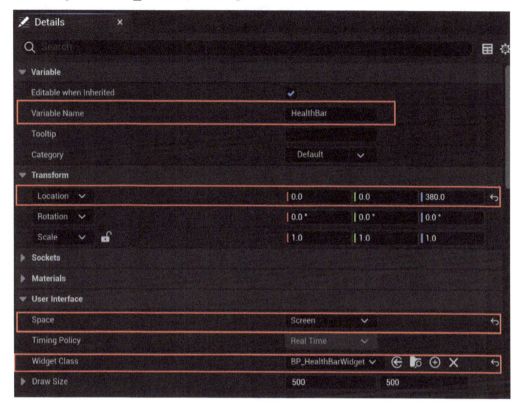

Figure 11.26 – The HealthBar widget component's settings

The last thing we must do is let the **Game Over** window display **You Win!** or **You Lose!** To do so, open **BP_GameOverWidget** and add the Blueprint graph shown here:

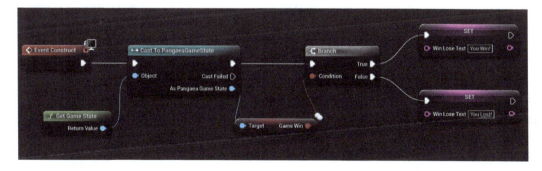

Figure 11.27 – Setting the Game Over result text graph

The blueprint checks the `ApangaeaGa,eState::GameWin` flag – if it is **True**, the **WinLose** text box, **You Win!**, is set; otherwise, **You Lose!** is set. Once a base tower is destroyed, the game should be paused and show the **Game Over** window with the **You Win!** text:

Figure 11.28 – The Game Over screen when the players win

With the game now reaching basic completion, it is crucial to allocate time to iteratively refactoring and refining both the game project and source code to enhance its overall quality. This practice aligns with the real game development process, which is required by the end of prototyping and adding new features.

Summary

In this chapter, you started by designing the basic game flow for *Pangaea* and then created three UI widgets called **BP_LobbyWidget**, **BP_HUDWidget**, and **BP_GameOverWidget**. You also learned how to use Unreal's `OpenLevel` function to start a listen server, join a game as a client, and travel back to the lobby. Based on knowing that a game instance always exists on the client side, you created the networking member functions for `APangaeaGameInstance`. By clicking buttons on the user interfaces, players can play the online game by choosing to host a game or join other people's game sessions and even exit the current game to go back to the lobby.

To add something fun to this game, you added some conditions for players winning or losing the game. The `Timer` value was used to determine whether the game ends due to a timeout. You also finished implementing **DefenseTower** by allowing designers to designate any defense tower to be the base tower so that once a base tower is destroyed, the players win the game.

The next chapter will show you how to package the project for publication and will include some useful recommendations on importing high-quality assets to polish the game using console commands and recovering corrupted projects.

12

Polishing and Packaging the Game

Welcome to the last chapter of this book. In the preceding chapters, we worked together and crafted a basically playable *Pangaea* game. The current stage beckons us to focus on polishing, packaging, and presenting the immersive *Pangaea* game to players.

We will discuss and explore a range of potential approaches to polishing and improving the game's quality. This will include incorporating high-quality assets, fixing bugs, and leveraging the engine's profiling tools to improve performance.

Some useful Unreal Engine console commands will be introduced so that you can use them to start the standalone game with preferred settings.

We will also outline the essential settings required for packaging the *Pangaea* game, then we will follow the steps to package and generate a standalone game for Windows. By providing this guidance, we aim to assist you in successfully preparing the game for distribution.

This chapter includes the following sections:

- Polishing the game
- Using Unreal Engine console commands
- Packaging the game
- What to do next

Technical requirements

The code for this chapter can be found at https://github.com/PacktPublishing/Unreal-Engine-5-Game-Development-with-C-Scripting/tree/main/Chapter12.

Polishing the game

Having completed the preceding 11 chapters, we have now arrived at a significant milestone where we possess a playable *Pangaea* game prototype that serves as a foundation for further development. Our next task, polishing the game, involves the following three aspects:

- Importing and using more game content
- Fixing bugs
- Profiling and improving the performance of the game

Let's get started.

Importing and using high-quality game assets

To enhance the visual quality and create a more immersive player experience, it is possible to incorporate additional assets into the game during the polishing phase. These assets may include high-quality artwork, audio clips, videos, and more. The acquisition of these art assets depends on the specific project requirements and available budget.

Certain types of assets can be used to polish the game, such as characters, structures, props, items, animations, particle effects, sound effects, music clips, and videos.

Of course, we can also create new content based on the new assets; for example, we can design new game levels with new terrains, plants, and buildings.

There are mainly two ways to obtain more assets for your games, as outlined here:

- **Epic Games Marketplace** provides a convenient way to acquire free assets or purchase third-party-developed assets (see *Figure 12.1*). To access the marketplace, simply open the **Epic Games Launcher** and select **Marketplace** from the top horizontal menu:

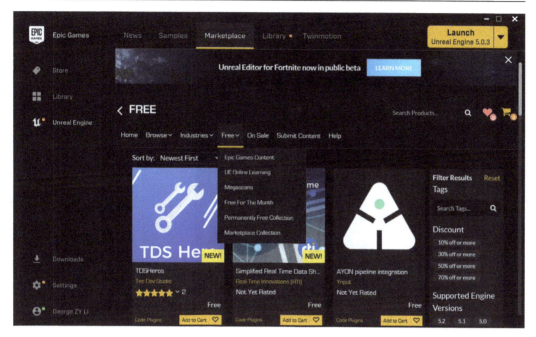

Figure 12.1 – Epic Games Marketplace

- One other way to obtain game assets is to hire designers and artists to develop unique game assets that will be used for your game in production. This investment needs to be carefully evaluated and well planned, depending on your development budget.

To showcase how high-quality assets can enhance the game's visual immersion, we can acquire the **Infinity Blade: Grass Lands** asset (see *Figure 12.2*) from the Epic Games Marketplace and add it to the *Pangaea* project:

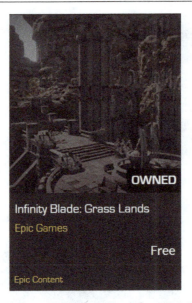

Figure 12.2 – Adding Infinity Blade: Grass Lands from Epic Games Marketplace

Subsequently, we can develop a new version of the **Topdown** game level by strategically placing enemies and towers within the environment (see *Figure 12.3*):

Figure 12.3 – Playing Pangaea in the new Topdown level

While delivering an enjoyable gameplay experience, the presence of bugs can introduce frustration if they are not promptly addressed. Therefore, we should always pay attention to bugs.

Fixing bugs

Bug fixing is an essential part of the development life cycle aimed at improving the functionality and reliability of the game code. Bug fixing is an iterative process that needs developers to repeatedly troubleshoot, identify, and fix bugs and eventually deliver a game that meets the expected quality.

Code review

Code review for bug fixes is a process where developers examine and evaluate code and address bugs. This process helps to ensure the correctness of logic and algorithms and adherence to coding standards, as well as identify potential issues or improvements. It is usually a multiple team members' collaboration practice.

Quality assurance

Quality assurance (**QA**) activities are usually carried out by QA teams that systematically test and inspect games to ensure that products or services meet specified requirements and quality standards. A found bug is reported by the QA tester and will be fixed by the developer. The process helps to enhance the overall quality of the product.

In addition to addressing bug fixes, optimizations play a vital role in enhancing the overall gameplay experience for players. The optimization process often entails the implementation of profiling practices.

Profiling and optimization

Profiling and optimization are essential aspects of game development. The main purpose of profiling is to identify CPU and GPU performance bottlenecks, memory usage, memory leaks, content and package-size issues, and so on. Optimization techniques vary depending on the detected issues and may involve code, algorithm, data structure, asset, and memory management optimizations.

Besides code analysis, profiling usually relies on a set of profiling tools. We are not going to look further into this topic in this book, but here is a list of some profiling tools:

- **Unreal Insights** is a set of profilers that assist developers to investigate performance, memory, networking, and UI details. Here is the link to the documentation page: `https://docs.unrealengine.com/5.0/en-US/unreal-insights-in-unreal-engine/`.

- The **Built-in** profiler tools, such as **GPU Visualizer** (press *Ctrl + Shift + ,* to open the **GPU Visualizer** window in the editor). Please refer to the official documentation for more information (`https://docs.unrealengine.com/5.0/en-US/testing-and-optimizing-your-content/`).

- Using the viewport **View Modes**, such as the **Wireframe** (*Alt + 2*), **Light Complexity** (*Alt + 6*), **Shader Complexity** views (*Alt + 7*), and so on.

Unreal Engine also offers a valuable arsenal of console commands that can be utilized in testing and tuning games. Let's explore the concept of the console command system and acquaint ourselves with a selection of fundamental commands.

Using Unreal Engine console commands

Unreal Engine allows testers to use console commands to interact with the game while playing. Console commands are widely used to change game settings, view the game status, manipulate game attributes, change game behaviors, tune game parameters, and so on.

Exploring modes and console commands

A pop-up UI called the **Debug** menu is usually developed to link UI actions (buttons, input boxes, and selections) to certain console commands. Testers can use the **Debug** menu to quickly jump into a particular game state. If you want to learn more about the **Debug** menu, you can visit `https://en.wikipedia.org/wiki/Debug_menu`.

Please be aware of the prerequisite for using console commands in standalone games, which requires games to be packaged as **Development** builds instead of **Shipping** builds.

To use console commands while playing the game, press the tilde key (~) on your keyboard to toggle between showing the small console, showing the large console, and hiding the console. Please be aware that the command console may show in slightly different places depending on which play mode (**Debug** or **Simulate**) you are currently using.

The compact **Debug** mode command console usually appears in the viewport of the engine editor:

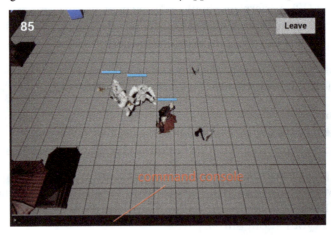

Figure 12.4 – The compact command console when it is in the Debug mode

The compact **Simulate** mode command console usually appears in seperate **Play In Editor** (PIE) windows:

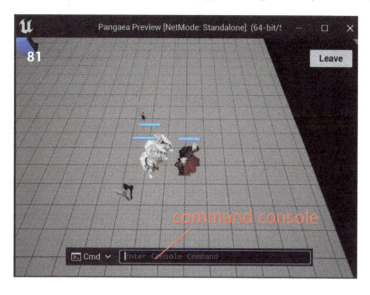

Figure 12.5 – The compact command console when it is in Simulate mode

The expansive command console occupies a big portion of the game screen, providing an extensive display of command history and previous command responses:

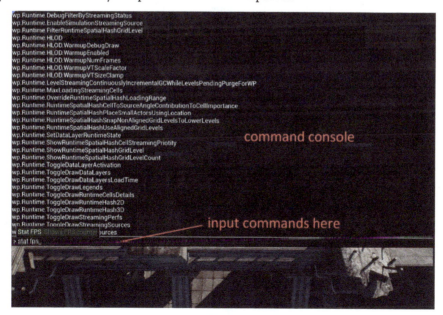

Figure 12.6 – The expansive command console

While it is not within the scope of this book to delve into every console command, we will focus on introducing a curated selection of useful commands listed in the following table to enhance your understanding:

Command	Description	Example(s)
`Exit`	Exits the game	`>exit`
`Stat FPS`	Toggles the visibility of the FPS	`>stat fps` (see *Figure 12.8*)
`Stat Unit`	Toggles the visibility of helpful performance info	`>stat unit` (see *Figure 12.9*)

`r.SetRes`	Sets the standalone game's screen resolution and the fullscreen or windowed mode	`>r.setres 1280x720w`
		
		(w stands for windowed)
		
		`>r.setres 1920x1080f`
		
		(f stands for fullscreen)

`Open [MapName]?Listen`	Starts the game on the given map as a Listen Server	`>open TopDown?listen` (Opens the TopDown level and starts the game as a Listen Server)
`Open [IP address] [:Port]`	Connects the game as a client to the remote server with the given IP address (for example, `127.0.0.1`) and the port number (for example, `7777`)	`>open 127.0.0.1:7777` (Connects to the `127.0.0.1:7777` server as a game client. The game map will be the same as the server's current map.)
`DumpConsoleCommands`	Prints out all console commands on the console	`>dumpconslecommands`

Figure 12.7 – Some Unreal Engine console commands

The following screenshots show the use of the `stat.fps` and `stat.unit` commands to enable the display of the FPS and basic performance information in the top-right corner of the game window.

When displaying the frame rate with `stat.fps`, the frame interval time is also presented beneath it:

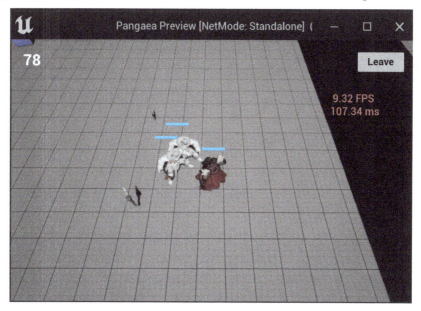

Figure 12.8 – Using the stat.fps command to display the current FPS

The `stat unit` command displays the CPU and GPU time spent on the current frame, as well as some of the current game performance information:

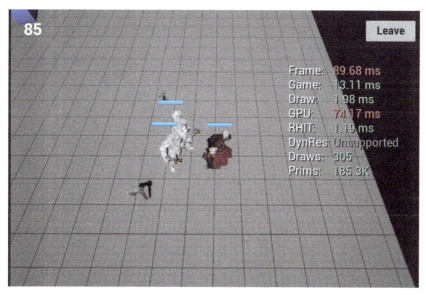

Figure 12.9 – Using the stat.unit command to display the current performance state

For further exploration of Unreal's console commands, you can visit the suggested websites or conduct online searches to access specific command documentation and related resources:

- `https://docs.unrealengine.com/5.2/en-US/stat-commands-in-unreal-engine/`

- `https://docs.unrealengine.com/5.2/en-US/audio-console-commands-in-unreal-engine/`

- `https://docs.unrealengine.com/5.2/en-US/console-commands-for-network-debugging-in-unreal-engine/`

- `https://pongrit.github.io/`

You can also utilize the dumpconsolecommands command to display a comprehensive list of all available commands directly in the console:

Figure 12.10 – Viewing the command list on the command console screen

Apart from entering console commands manually, Unreal also offers an API that allows the execution of console commands in Blueprint and C++. Let's delve into writing the C++ code to execute console commands.

Executing console commands in C++

In Unreal Engine C++, you can execute console commands by calling the `APlayer Controller::ConsoleCommand` function.

Here is the syntax of the `ConsoleCommand` function:

```
bool ConsoleCommand(const FString& Command, bool bWriteToLog);
```

Let's break down this syntax:

- Command parameter: The console command that will be executed

- bWriteToLog parameter: Determines whether the execution of this command will be logged (true) or not (false)

- The return value of the function is a bool value that indicates whether the command execution succeeded (true) or failed (false)

Refer to the following code example to learn how to execute console commands programmatically in C++, and make sure that this sample code is part of an UObject or AActor subclass's member function:

```cpp
#include "Engine/World.h"
#include "GameFramework/PlayerController.h"

// ...
auto World = GetWorld();   //Get the world

if (World)
{
    auto PlayerController = World->GetFirstPlayerController();

    if (PlayerController)
    {
        FString Command = TEXT("stat fps");
        PlayerController->ConsoleCommand(Command, true);
    }
}
```

In the previous code snippet, you first obtain a reference to the current UWorld object using the GetWorld() function. Then, you retrieve the first player controller using the world's GetFirstPlayerController() member function. Finally, you execute a console command (for example, stat fps) using the ConsoleCommand() function of the APlayerController class.

The last thing we want to do in this book is to package the *Pangaea* game. Let's explore the packaging process steps.

Packaging the game

During the development process, we extensively played the *Pangaea* game within Unreal Editor. However, our goal is to make the game accessible to players without requiring them to install Unreal Engine. Luckily, Unreal Engine provides a convenient packaging feature that allows us to create standalone install packages for various platforms. This means we can distribute the *Pangaea* game as a separate application for Windows, macOS, iOS, Android, and more. To demonstrate this, we will create a *Pangaea* Windows game installation package.

Prior to the packaging process for the game, certain project settings need to be configured.

Configuring the project settings for packaging

To successfully package a game, as the bare minimum, there are two tasks you need to complete, as follows:

- The first task involves setting the project defaults, which entails designating the default game mode, player pawn, and so on
- The second task involves including the necessary game levels within the generated installation package, ensuring that the game levels are present for the gameplay

Let's start setting the project defaults in the project settings.

Setting the project defaults

To work on the settings, choose **Edit | Project Settings…** from the editor's main menu. Then, from the **Project Settings** window, find and select **Maps & Modes** under the **Project** group on the **All Settings** panel. Now, choose the following settings:

- **Default Game Mode**: `PangaeaGameMode`
- **Default Pawn Class**: `BP_PlayerAvatar`
- **Player Controller Class**: `BP_TopdownPlayerController`
- **Game State Class**: `PangaeaGameState`
- **Game Default Map**: `LobbyMap`
- **Game Instance Class**: `PangaeaGameInstance`

We can see these settings in the following screenshot:

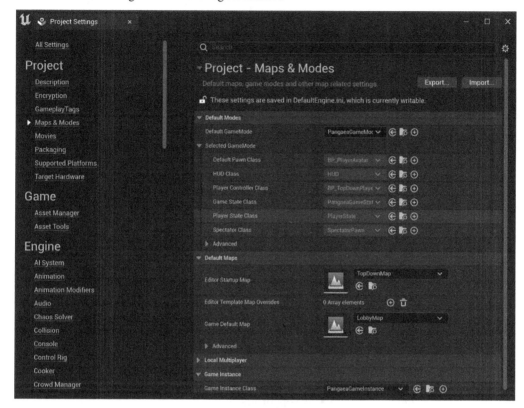

Figure 12.11 – Project – Maps & Modes settings

Next, we want to add the required game levels to the list of included maps.

Including the game levels in the built package

In the **Project Settings** window, find and select **Packaging** under the **Project** group. Then, click the **Add Element** button (denoted by the + symbol) to add LobbyMap and TopDownMap to the **List of maps to include in a packaged build** setting:

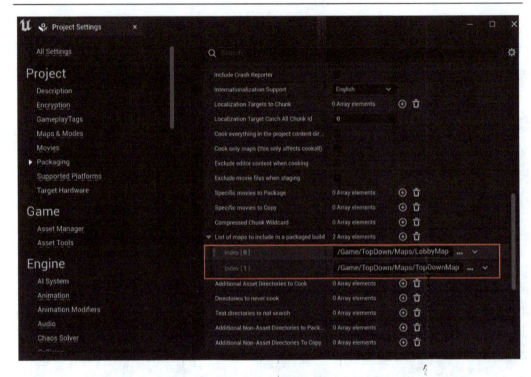

Figure 12.12 – Adding game maps that will be included in the package build

The project settings are now prepared for packaging the project, but if we desire the built game to run in the 1,280x720 window mode, this is the opportune moment to utilize the `r.setres` console command to accomplish it.

Making the build a windowed game

By default, the packaged standalone build of the game is launched in fullscreen mode. However, it is more convenient to test the multiplayer functionality by running two instances of *Pangaea* in separate windows on your machine.

To facilitate this, you can configure the game to start as a windowed application with a resolution of 1,280x720 and ensure that the mouse cursor is visible. To achieve this, a minor modification can be made to the level blueprint of the LobbyMap level blueprint, as follows:

1. Execute the console command to start the game in the 1,280x720 window mode.

2. Check the **Show Mouse Cursor** checkbox in the **SET** node:

Figure 12.13 – Modifying the LobbyMap level blueprint

We just accomplished the task by utilizing the LobbyMap level blueprint to ensure we start the game with the required resolution and in window mode. One more thing to improve is to remove the hardcoded path for finding the BP_Fireball and BP_Hammer blueprint classes for spawning the actors.

We just completed a task that allowed us to ensure that the game starts with the required resolution and in window mode. One further improvement is to eliminate the hardcoded path used to locate the BP_Fireball and BP_Hammer blueprint classes for spawning actors.

Avoiding the hardcoded path for finding content

Within the game's C++ source code, we utilized ConstructorHelpers::FObjectFinder to obtain references to the BP_Fireball and BP_Hammer classes for spawning fireball and hammer instances. However, this approach employs hardcoded paths to the assets, which may lead to runtime errors when the target asset is missing or if the specified path cannot be located.

The following code snippets depict the original implementation of the ADefenseTower and AEnemy classes that we aim to replace.

Here is DefenseTower.cpp:

```
static ConstructorHelpers::FObjectFinder<UBlueprint> blueprint_
finder(TEXT("Blueprint'/Game/TopDown/Blueprints/BP_Fireball.BP_
Fireball'"));
_FireballClass = (UClass*)blueprint_finder.Object->GeneratedClass;
```

And here is Enemy.cpp:

```
static ConstructorHelpers::FObjectFinder<UBlueprint> blueprint_
finder(TEXT("Blueprint'/Game/TopDown/Blueprints/BP_Hammer.BP_
Hammer'"));
_WeaponClass = (UClass*)blueprint_finder.Object->GeneratedClass;
```

To improve this code, we want to define _FireballClass and _WeaponClass as UPROPERTY variables so that we can select and specify **Blueprint** classes in the editor.

Let's start by modifying the DefenseTower class:

1. In Enemy.h, change the _FireballClass variable type to TSubClassOf<AProjectile>. Then, move the variable definition code to the public section. To maintain the code standard, we can change the variable name to FireballClass (removing the heading underscore, _):

    ```
    public:
        UPROPERTY(EditAnywhere, Category = "Tower Params")
        TSubclassOf<AProjectile> FireballClass;
    ```

2. In Enemy.cpp, remove the following two lines of code:

    ```
    static ConstructorHelpers::FObjectFinder<UBlueprint>
        blueprint_finder(TEXT("Blueprint'/Game/TopDown/Blueprints/
        BP_Fireball.BP_Fireball'"));
    FireballClass = (UClass*)blueprint_finder.Object-
    >GeneratedClass;
    ```

3. Change the parameter name from _FireballClass to FireballClass when calling the SpawnOrGetFireball function:

    ```
    void ADefenseTower::Fire()
    {
    auto fireball = _PangaeaGameMode->SpawnOrGetFireball(
                            FireballClass);

        ...

    }
    ```

4. Select BP_Fireball for the FireballClass field for BP_DefenseTower in the editor. Then, do the following:

    I.   Compile and reopen the project in the Unreal Editor.

    II. Open **BP_DefenseTower** from the **Content | Topdown | Blueprints** folder.

III.   Select BP_Fireball for the **Fireball Class** field:

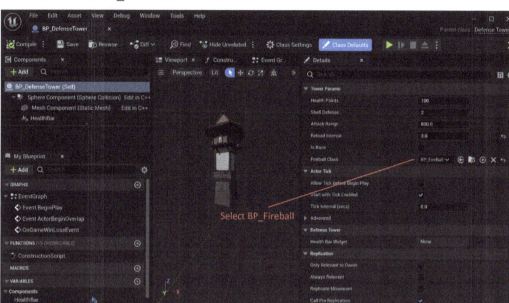

Figure 12.14 – Selecting BP_Fireball for the FireballClass field on the blueprint editor

5.   Compile and save BP_DefenseTower.

Similarly, we can apply the aforementioned steps to make modifications to the AEnemy class. In Enemy.h, define the WeaponClass variable as a subclass of AWeapon:

```
public:
  UPROPERTY(EditAnywhere)
  TSubclassOf<AWeapon> WeaponClass;
```

Then, comment or remove the two asset finder lines of code from the constructor of AEnemy in Enemy.cpp:

```
AEnemy::AEnemy()
{
  ...
  //static ConstructorHelpers::FObjectFinder<UBlueprint>
    blueprint_finder(TEXT("Blueprint'/Game/TopDown/Blueprints/
    BP_Hammer.BP_Hammer'"));
  //_WeaponClass = (UClass*)blueprint_finder.Object->GeneratedClass;
}
```

```
void AEnemy::BeginPlay()
{
    Super::BeginPlay();
_Weapon = Cast<AWeapon>(
            GetWorld()->SpawnActor(WeaponClass));
    ...
}
```

We are now ready to package the game build.

Packaging the project

To package the project, we can complete the following steps:

1. Click **Platforms** on the editor's toolbar. Then, select **Windows**.

2. Next, choose one of the packaging configuration types (see *Figure 12.15*):

    I.   **DebugGame**: When packaging with this configuration, the game engine code is optimized, whereas the game code is debuggable without optimizations

    II.   **Development**: This is the engine's default compiling configuration, which only optimizes the most time-consuming engine and game code but leaves the other code unoptimized and debuggable

    III.   **Shipping**: When packaging with this configuration, all debug symbols—including logs, status, profiling data, and so on—are stripped off, and the project is fully optimized for the best game performance

3. Click the **Package Project** item on the second-layer menu.

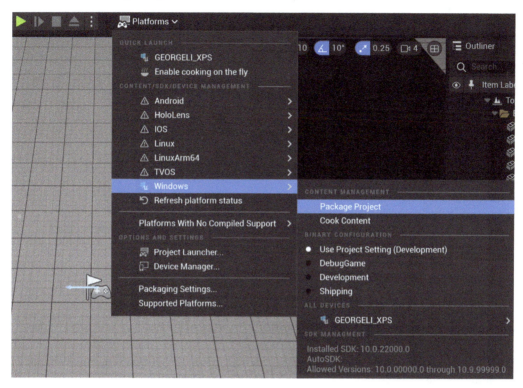

Figure 12.15 – Starting to pack the game

4. Select the target folder for where the packaged build will be saved (see *Figure 12.16*). Once the **Select Folder** button is pressed, Unreal will start the packaging process:

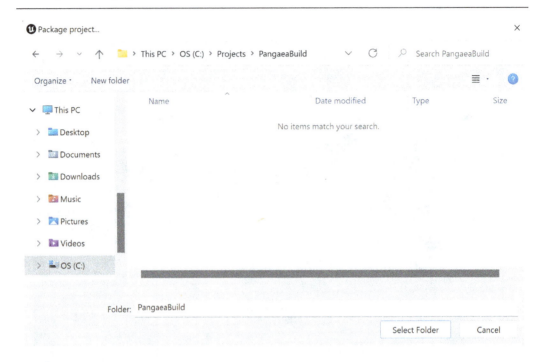

Figure 12.16 – Selecting the target folder (for example, C:\Projects\PangaeaBuild) to save the game build

5.   While packaging, you can view the progress and the logs in the **Output Log** window located at the bottom of the editor:

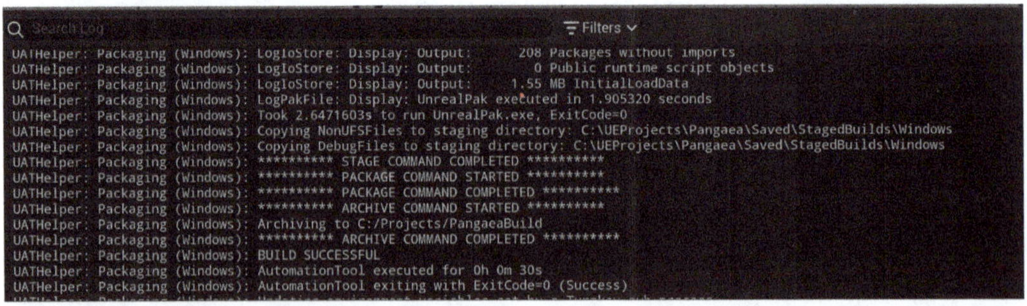

Figure 12.17 – Viewing packaging progress logs in the Output Log window

6. Once the packaging process is complete, you can locate the packaged files and their corresponding subfolders in the designated target folder (see *Figure 12.18*):

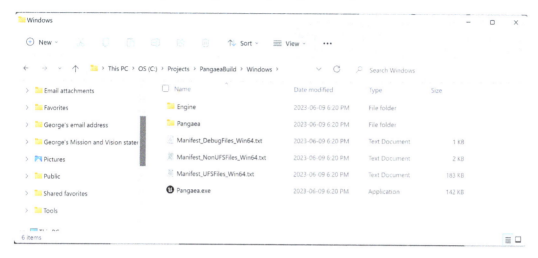

Figure 12.18 – The generated package build files

7. Then, to launch the standalone game, simply double-click on the `Pangaea.exe` executable.

Congratulations on acquiring the playable multiplayer *Pangaea* game build, which you can now copy and distribute to share the game with others!

What to do next

This book is structured to provide comprehensive insights and techniques essential for creating and developing *Pangaea*, a multiplayer game in the top-down genre. It covers a range of topics, including fundamental C++ coding skills, the Unreal `Actor` class and its subclasses, animation control, player interactions, collision handlers, UI, and multiplayer game fundamentals. However, the book does not delve extensively into each area and could not cover all aspects of techniques for game development as there is a vast and intricate body of knowledge to explore. Becoming an advanced expert in Unreal C++ game development requires continuous learning, practice, and experience accumulation to further refine your skills.

The key to success lies in consistent practice and dedication. You can choose to embark on a new project from scratch or enhance the existing *Pangaea* game by gradually incorporating additional features and exploring advanced technologies. Aim to gain familiarity with APIs, C++ syntax, the Visual Studio IDE, and the UE5 development environment. Cultivate your ability to analyze and solve problems, and continuously refactor, refine, profile, and optimize your ongoing development efforts using an iterative process. This will enable you to broaden your knowledge base and explore different directions within various areas.

Engage in practices such as refactoring, refining, and adding new features, as well as digging into certain areas. When seeking to expand your understanding of specific technologies, the Unreal Engine API documentation (`https://docs.unrealengine.com/5.0/en-US/API/`) and the abundance of online tutorials serve as excellent references.

Summary

What you just learned in the chapter was ways of polishing the *Pangaea* game by importing new assets, fixing bugs, profiling to improve performance, and using console commands to manipulate and tune the game. You also followed instructional steps to configure the packaging settings and generated a Windows executable game for distribution. The last section of this book also provided my suggestions for your further learning based on the knowledge and skill set you gained from this book.

I hope this book can help to accelerate your learning experience and be the starting point for your Unreal Engine C++ game development career. I wish you all the best with your future professional game development career and I hope you make your dream games come true!

Index

A

B

E

F

G

Packtpub.com

Subscribe to our online digital library for full access to over 7,000 books and videos, as well as industry leading tools to help you plan your personal development and advance your career. For more information, please visit our website.

Why subscribe?

- Spend less time learning and more time coding with practical eBooks and Videos from over 4,000 industry professionals

- Improve your learning with Skill Plans built especially for you

- Get a free eBook or video every month

- Fully searchable for easy access to vital information

- Copy and paste, print, and bookmark content

Did you know that Packt offers eBook versions of every book published, with PDF and ePub files available? You can upgrade to the eBook version at packtpub.com and as a print book customer, you are entitled to a discount on the eBook copy. Get in touch with us at customercare@packtpub.com for more details.

At www.packtpub.com, you can also read a collection of free technical articles, sign up for a range of free newsletters, and receive exclusive discounts and offers on Packt books and eBooks.

Other Books You May Enjoy

If you enjoyed this book, you may be interested in these other books by Packt:

Blueprints Visual Scripting for Unreal Engine 5

Marcos Romero | Brenden Sewell

ISBN: 9781801811583

- Understand programming concepts in Blueprints.
- Create prototypes and iterate new game mechanics rapidly.
- Build user interface elements and interactive menus.
- Use advanced Blueprint nodes to manage the complexity of a game.
- Explore all the features of the Blueprint editor, such as the Components tab, Viewport, and Event Graph.
- Get to grips with OOP concepts and explore the Gameplay Framework.

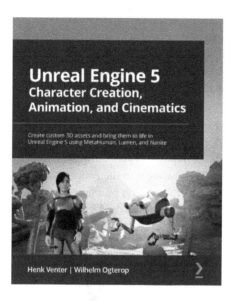

Unreal Engine 5 Character Creation, Animation, and Cinematics

Wilhelm Ogterop | Henk Venter

ISBN: 9781801812443

- Create, customize, and use a MetaHuman in a cinematic scene in UE5.

- Model and texture custom 3D assets for your movie using Blender and Quixel Mixer.

- Use Nanite with Quixel Megascans assets to build 3D movie sets.

- Rig and animate characters and 3D assets inside UE5 using Control Rig tools.

- Combine your 3D assets in Sequencer, include the final effects, and render out a high-quality movie scene.

- Light your 3D movie set using Lumen lighting in UE5.

Packt is searching for authors like you

If you're interested in becoming an author for Packt, please visit `authors.packtpub.com` and apply today. We have worked with thousands of developers and tech professionals, just like you, to help them share their insight with the global tech community. You can make a general application, apply for a specific hot topic that we are recruiting an author for, or submit your own idea.

Share Your Thoughts

Now you've finished *Unreal Engine 5 Game Development with C++ Scripting*, we'd love to hear your thoughts! Scan the QR code below to go straight to the Amazon review page for this book and share your feedback or leave a review on the site that you purchased it from.

`https://packt.link/r/1-804-61393-2`

Your review is important to us and the tech community and will help us make sure we're delivering excellent quality content.

Download a free PDF copy of this book

Thanks for purchasing this book!

Do you like to read on the go but are unable to carry your print books everywhere?

Is your eBook purchase not compatible with the device of your choice?

Don't worry, now with every Packt book you get a DRM-free PDF version of that book at no cost.

Read anywhere, any place, on any device. Search, copy, and paste code from your favorite technical books directly into your application.

The perks don't stop there, you can get exclusive access to discounts, newsletters, and great free content in your inbox daily

Follow these simple steps to get the benefits:

1. Scan the QR code or visit the link below

https://packt.link/free-ebook/9781804613931

2. Submit your proof of purchase
3. That's it! We'll send your free PDF and other benefits to your email directly

www.ingramcontent.com/pod-product-compliance
Lightning Source LLC
Chambersburg PA
CBHW080610060326
40690CB00021B/4646